Chemistry
Through the Language
Barrier

Chemistry
Through the Language
Barrier

How to Scan Chemical Articles in Foreign Languages
with Emphasis on Russian and Japanese

E. Emmet Reid

THE JOHNS HOPKINS PRESS
BALTIMORE AND LONDON

Copyright © 1970 by The Johns Hopkins Press
All rights reserved
Manufactured in the United States of America

The Johns Hopkins Press, Baltimore, Maryland 21218
The Johns Hopkins Press Ltd., London

Library of Congress Catalog Card Number 75–112360

Standard Book Number 8018–1109–0

Contents

Foreword

"Can *I* play the piano?
Why, I don't know — I never tried."
(MARK TWAIN)

"I never knew I could
read Czech until I tried."
(E. EMMET REID)

In his long and fruitful career, my revered colleague Dr. Reid, the oldest Hopkins professor, the inveterate researcher, has gone over literally acres of chemical papers in German and French and related languages. In the past ten years he has tackled articles in languages he did not know, and to his amazement discovered he could get information out of them, too! He knows, therefore, whereof he speaks. His straightforward philosophy in the matter of scientific reading is peculiarly instructive and timely nowadays, when the language requirements for the Ph.D. degree are being debated, and debased, in universities throughout the land. His advice is directed at chemists, but any scientist, engineer, or mathematician can benefit from it.

The fundamental truth is this: if you are investigating any question, you want to know *all* that has been said on it. More important yet, you want to find out what is being written about it *now* and to get the information with minimum delay. You are so eager indeed that, when the current number of a journal comes in, you immediately grab it and start searching through it for any article that might possibly be relevant to your quest. You dutifully scan the abstracts of the works published in the whole world, hopefully looking for a contribution to your specialized field. What are you to do when you discover that the paper you so much want to read is in Czech or in some other foreign tongue? Don't hesitate. Do what Dr. Reid tells you — try.

Every science, chemistry in particular, has developed its own symbolism which is internationally understood. Numbers, chemical formulae and equations will look familiar even in a Japanese text. Technical terms, too, are international and will be recognized. Add the long lists of cognate words and of words that languages borrow from one another, and you have the foundation on which to build. Dr. Reid says, "I am no linguist: German and French are the only languages I studied; in other languages I just picked up the information." He goes on, "The more you know about your field, the easier it will be to find your way through the foreign paper." In other words (if I may be allowed the para-phrase): if you want to read Czech, study your chemistry! Now, what chemistry department could possibly take exception to such excellent advice? You must be a better chemist to read Czech to be a better chemist to read Russian to be a better chemist. . . . A perfectly "virtuous" circle that brings reward either way!

Dr. Reid does not pretend that the task he is assigning to you is an easy one, nor does he claim that you can read a novel or a poem in any language you care to pick. All he guarantees is that even the little bits you can gather from looking at a foreign chemical paper in your field of interest are well worth your trouble. Perhaps you will miss the fine points, but you may get the gist. At the very least you can ascertain whether the paper is of sufficient interest to you to justify your ordering a professional translation.

Contrast this practical approach of a veteran research chemist with the half-hearted and musty university requirements, which misplace the emphasis by making you *pass reading tests*, purporting to show that you are "able to read" German, French, Russian, or whatever, instead of scrapping the tests and show-ing you how information can be extracted from any foreign paper. Consider the fallacy of the theory that "it is better to study *one* foreign language thoroughly than to attempt to read two." For the specific task at hand, which is to comb the chemical literature of the world with a view to finding new information on a definite topic, it is clear that a smattering of ten to twenty languages will be more useful than the most polished command of French. The suggestion has also been proffered that FORTRAN, or some other computer jargon, should be an ac-ceptable substitute for French, German, or Russian, as a major publication language fulfilling Ph.D. requirements. This proposal is pure double-talk: What scientific journal is published in FORTRAN? FORTRAN, today, is as indispensable as the vector notation, but neither is a language in the present context. Dr. Reid's book, simple, straightforward, not in the least polemical, unerringly though unwittingly disposes of all our sophistry and brings us to our senses.

When it comes to languages as languages, Dr. Reid is a humanist of great sensitivity. "I love languages," he admits, "because they tell you so much about the people who speak them." I share Dr. Reid's fascination. Of course, languages can become great cultural assets if one is willing to study them. But what better introduction to a new idiom can there be than its firsthand inspection and observation, as recommended by Dr. Reid? Looking at the facts may lure us into venturing more deeply into the new language, studying its constructions, its grammar, its idiomatic expressions, trying to read its literature, wanting to speak it and hear it spoken, striving to understand the people who use it — and thereby becoming better citizens of a more peaceful world.

J. D. H. DONNAY

Baltimore, Maryland

Preface and Acknowledgments

Chemists have quite different approaches to the problem of getting information from articles in foreign languages.

The pessimist says: The journal is not in our library. The optimist replies: There are two hundred libraries in the United States that furnish photostats. I am sure I can get it.

The pessimist says: It is in Polish, and I can't read it. The optimist replies: I do not know a word of Polish but if it is on a subject with which I am familiar I am sure I can find in it enough familiar formulae, numerals, chemical names and technical words to tip me off as to what goes on. To get the information may require ingenuity and effort but no more than other problems which I face.

The pessimist says: This Russian article is impossible. The optimist replies: The trouble with it is the strange alphabet. I can't afford to let a string of letters stand as a barrier between me and what I need to know.

Samson with his long hair cut off was no stronger than other men. Russian, with its strange alphabet shaved off, is no more difficult than other languages.

The assistance of more than a score of chemists on texts in their native tongues is gratefully acknowledged. In addition to those whose names appear on sections in the text, valuable assistance has been received from Dr. George F. Bulbenko and the following: A. M. Alvarado, Irene E. Berck, Victor G. Fourman, Friedrich W. Hoffmann, Werner Jacobson, Edward A. Metcalf, Ryozo Motoyama, and John R. Sampey. Dr. J. S. Fok helped with revisions of the Japanese chapter.

These assistants deserve special thanks: Dawn Obrecht, David McGranaghan, Greg Tawes, Gerald Yee, and Louis Fries.

E. EMMET REID

Baltimore, Maryland

Chemistry
Through the Language
Barrier

1

Direct Approach to Chemical Literature
in Foreign Languages

———

Scientists and engineers used to worry about the sound barrier, but then they learned to ignore it — and now there are some airplanes that fly at more than three times the speed of sound. Chemists also have trouble with a barrier, the language barrier. But they are a hardy breed, accustomed to overcoming difficulties; why should they be halted by a language barrier?

SEARCHING — RATHER THAN READING —
CHEMICAL ARTICLES

It is seldom that a chemist reads chemical articles for entertainment. Present-day articles are so specialized that a chemist finds it hard going when he wanders out of his own narrow field. A dye chemist would get little from an article on progesterone. In a recent issue of the *Journal of the American Chemical Society* there are, in two hundred and ninety-seven pages, seventy-six articles on almost as many different subjects. Few chemists would or could read more than a few of these. But when it comes to his own specialized research, the chemist needs and wants to read all available articles which might have a bearing on his problem. Usually these cover much the same ground, but each presents something new. A particular article, like an installment in a continued story, is to be understood in its relation to other articles in the series.

1

However, the story that the chemist follows has a peculiar twist: The installments might be developed in widely separated geographical sites and reported in many different languages. What should he do when he comes across an article printed in an unfamiliar language and suspected of containing desirable information? He should disregard the language barrier, use the "direct approach," and go after the information he wants. This book will show him how this can be done.

No pretense is made that extracting worthwhile information from a foreign language article is easy, particularly if it is Russian or Japanese. Transporting a ton of coal from the bottom to the top of a hill requires many foot-pounds of energy, regardless of mechanism. However, the man who has a bucket and shovel can do it, even if there is no conveyor belt to help him in his task.

Getting information out of a foreign article is a real task which requires effort. The direct approach is for those who have no better way. Its great virtue is that it can be used effectively and yield results even in languages with which the chemist has had no prior experience. What the chemist must know thoroughly is his chemistry — and then be willing to use his ingenuity.

DISTRIBUTION OF CHEMICAL ARTICLES
ACCORDING TO LANGUAGE

There was a time when most chemical articles were published in German, with a smaller number in English and French. The chemist who knew these three languages could keep up with his field quite comfortably — but this time has passed. The explosively rapid expansion of chemical research and the proliferation of chemical literature makes the present-day chemist feel like a lone bookworm in the Library of Congress. However, what really worries him is that almost half the chemical articles that are not in English are in Russian.

Six languages are used in over ninety percent of all current chemical publications. In order of frequency of the articles, these languages are:

	1965	1969
English	52.0%	55.9%
Russian	20.0	22.6
German	9.8	7.5
Japanese	4.0	3.3
French	5.1	4.6
Italian	1.9	1.5

Of the remainder, an appreciable number appears in about a dozen other languages, while a few articles are scattered among some thirty more languages.

If the total number of chemical articles is classified according to the alphabet used, about seventy percent are written in the Latin alphabet, twenty percent in the Russian alphabet, and almost all the rest in the Japanese alphabet. This is the situation with which the chemist has to deal: he needs facility in reading Russian, German, and French, and ability to retrieve occasional items out of many other languages.

For this formidable task, the chemist is not entirely unprepared. In his laboratory training he has learned to identify a compound by observing a few of its properties. He can apply the same technique to examination of foreign chemical articles. When he finds some familiar symbols and words, he can figure out what process is described and compare it with what he already knows about the subject. With this beginning, he can continue by whatever means he finds necessary.

This book is written to encourage and help the chemist to go directly to any chemical article that promises to bear on his field of specialization, rather than wait for a translation to be published. This is not startlingly new — it is just what experienced chemists have been doing all along. But the non-Latin alphabets present special difficulties which the chemist must find ways to overcome. For that reason Russian and Japanese will be given special consideration here.

ARTICLES WRITTEN IN FOREIGN ALPHABETS

RUSSIAN

We may well regret that Cyril did not carry to the Slavic tribes, along with Christianity, the Latin alphabet rather than the Greek, but he could not have anticipated the trouble this was to cause future chemists, comparatively few of whom were prepared to meet the upsurge of Russian publications. As they have not been able to avoid it, they must learn how to cope with it. The essential thing is to gain familiarity with the Russian alphabet, then to acquire a limited, specialized Russian vocabulary relating to a narrowed field of interest. The task is not easy, but it is not as hard as it first appears. The method here given is for those who have no better way.

We approach Russian in three stages: the efficacy of the direct approach is demonstrated first on a Polish article (written in the Latin alphabet); attention is called to the similarity of Polish and the transliterated Russian text; finally, encouragement and help is offered in learning the Russian alphabet and working with it. As the chemist encounters many Russian chemical articles, it is desirable for him to develop facility in the language as fast as possible.

JAPANESE

Japanese articles are an important part of chemical literature and cannot be overlooked. The chemist is encouraged to use the direct approach (with necessary modifications) in dealing with them, just as he uses it with articles in any other language. This turns out to be easier than it looks. A detailed discussion of direct approach to Japanese is given in Chapter 4. At this point, only the general principle is noted.

In their writings, the Japanese use three thousand Chinese ideographs and two alphabets, the Katakana and the Hirakana. The chemist must get what he can from formulae and numerals printed as in English and then transliterate what is given in the Katakana alphabet. In this manner he can identify many familiar English and German names of compounds and other technical words — the percentage can be as high as seventy-five percent. The remainder he can look up in a dictionary of transliterated Katakana words. At this stage, he is likely to have most of the story, but will need additional information from the ideographs to complete it. To deal with these, he uses a guessing game. He thinks up an English word that might fit in a certain place, then looks it up in an English-Japanese dictionary and matches the ideographs he finds with those in the article. If they do not match, he must guess again. No matter how tedious this method is, it can yield results — provided the chemist has a good knowledge of his subject matter. With each article he gains speed in transliterating and saves time by remembering ideographs.

INTERNATIONAL CHEMICAL LANGUAGE

SYMBOLS

Chemists all over the world communicate with one another in a written code consisting of symbols for the elements and numerals indicating their proportions. Symbols like CH_4, C_2H_4, C_6H_6, and $C_6H_5NO_2$, designating methane, ethylene, benzene, and nitrobenzene, are examples of this international code and are written, read, and understood by chemists of every nation on earth. Formulae using these symbols appear in all languages exactly the same as they are printed in the *Journal of the American Chemical Society*. Arabic numerals which are used for weights and measures, melting points, boiling points, densities, and the like are also parts of this international code.

INTERNATIONAL WORDS

If substances are identified by names rather than symbols, words vary somewhat from language to language, but many are still easily recognizable — e.g., ethylene (English), etilene (Italian), aetylen (Danish), Äthylen (German), etylen (Polish).

As chemical ideas and operations have spread from country to country, they have carried with them hundreds of technical words, such as "distill" and "crystallize." These may be somewhat altered but they are still identifiable; the chemist has little difficulty understanding "synthesis" whether it is spelled "synteze" (Polish) or "sintez" (Russian), or "reaction" whether it appears as "reakcja" (Polish) or "reazione" (Italian). Such words make up an important part of the international language of chemistry.[1]

SOME LIMITATIONS

The fact that the same systematic names of *many* organic compounds are used in all languages with only slight variations helps the chemist with *some* names (like "nitrophenol") but not with *all* names — there are always words like "Ameisensäure" (HCOOH), "Essigsäure" (CH$_3$COOH), and "Schwefelsäure" (H$_2$SO$_4$). Inserting formulae after such words would be of great service to the international readership and little trouble to the author, but the practice is seldom followed. Enlarging the international language by addition of a number of common words, such as "heating" and "cooling" would also help. But, instead of contemplating what would help, it is well to assess what the chemist has now.

EXAMPLES OF INTERNATIONAL WORDS FROM A SAMPLE TEXT

The list which follows gives the chemical names and international words found in translations into fifteen different languages of a section from Sabatier's "La catalyse en chimie organique."[2] The article from which this text is taken is noteworthy as one of the very first writings on catalytic hydrogenation. Attention is called to the fact that a number of chemical names and international words run through all versions. The chemist experienced in catalytic hydrogenation and familiar with one version would have little difficulty with the others. It is to be particularly noted that transliterated Russian words are practically the same as their Polish equivalents. Transliterated Japanese would also reveal these same, familiar-looking words.

[1] In Appendix D, nearly four hundred of these words known to be used in Russian are listed.

[2] Paul Sabatier, "La catalyse en chimie organique," *Encyclopédie de science chimique appliquée*, Edit. Béranger, vol. 3, p. 195.

Cross-Language Comparison of International Chemical Words
Found in Sample Text

French	Italian	Spanish	Romanian	Portuguese
acétylène	acetilene	acetileno	acetilenă	acetileno
hydrogène	idrogeno	hidrógeno	hidrogen	hidrogênio
platine	platino	platino	platin	platina
éthylène	etilene	etileno	etilenă	etileno
éthane	etano	etano	etan	etano
hydrocarbures	idrocarburi	hidrocarburos	hidrocarburi	hidrocarbonetos
nickel	nichel	niquel	nichel	níquel
produit	prodotto	producto	produs	produto
transformé	trasformato	transformado	transformată	transformado
proportion	proporzione	proporción	proportie	proporção
incandescence	incandescenza	incandescencia	incandescență	incandescência
analogue	analoga	——	——	——
réaction	reazione	reacción	reacție	reacçaõ
présence	presenza	presencia	prezența	presença
se combine	si combina	se combina	prin combinație	se combina
gaz	gas	gas	gas	gás
hydrogénation	idrogenazione	hidrogenación	hidrogenarea	hidrogenação
——	——	secundario	secundare	secundárias
formation	formazione	formacion	formarea	formação
contact	contatto	contacto	contact	——
——	——	——	descompus	decomposição
——	——	——	——	——
——	——	——	——	——
——	——	——	——	——
active	attiva	activo	activ	activo
——	——	——	——	——

German	Dutch	Swedish	Norwegian	Danish
Acetylen	acetyleen	acetylen	acetylen	acetylen
——	——	——	hydrogen	——
Platin	platinum	platina	platina	platin
Äthylen	ethyleen	etylen	etylen	ætylen
Äthan	ethaan	etan	etan	ætan
——	——	——	hydrokarboner	——
Nickel	nikkel	nickel	nikkel	nikkel
Produkt	product	produkt	produkt	produkt
——	getransformeerd	——	——	——
——	——	——	——	——
——	——	——	——	inkandescens
——	——	——	——	——
Reaktion	reactie	reaktion	reaksjon	reaktionen
——	——	——	——	——
——	——	——	——	——
Gas	gas	——	gass	——
Hydrierung	hydratie	hydrerar	hydrogenering	——
——	——	——	——	——
——	——	——	——	——
——	——	——	——	——
——	——	——	——	dekomponeret
——	temperatuur	temperaturer	temperatur	temperatur
quantitativ	——	——	——	——
Konzentration	——	——	——	——
——	actief	aktiv	aktiv	——
——	——	katalysatorn	katalysator	katalysator

Hungarian	Finnish	Russian	Polish	Czech
acetilén	asetyleeni	atsetilen	acetylen	acetylen
hidrogénnel	——	——	——	——
platina	platina	platinovoi	platyny	platinové
etilént	etyleeniä	etilen	etylen	ethylen
etánt	etaania	etan	etan	ethan
——	——	——	——	——
nélkül	nikkeliä	nikela	niklem	niklu
——	——	produktov	produktem	produkt
——	——	——	——	——
——	——	——	proporcji	——
——	——	——	——	——
——	——	analogicheskoi	——	——
reakció	rektio	reaktsiya	reakcja	reakce
——	——	——	——	——
——	——	——	——	——
gáz	——	gaza	gazu	——
——	——	hidrogenizatsiyi	——	hydrogenaci
——	——	——	——	——
——	——	——	——	——
——	——	——	——	——
——	——	——	——	——
——	——	——	temperaturze	——
——	——	——	——	——
——	——	——	——	——
——	——	aktivnost	aktywna	——
katalizálja	——	——	——	——

APPLICATION OF THE DIRECT APPROACH

In this book, texts chosen for analysis are taken from experimental sections of articles on the preparation of organic compounds. These descriptions are usually straightforward, give the names of the starting materials and the treatments to which they are subjected. They are eminently suitable for the chemist who wants to "get his foot in the door." Articles in physical chemistry would have algebraic equations instead of chemical formulae and determinations of orders of reactions in place of the percentage composition of compounds.

PROCEDURE

Examination of an article consists of a series of steps. The first is to learn from any clues the chemist may find whether the article really concerns his particular problem. Next, the chemist checks the names and relative amounts of the starting materials, so as to relate the process to one with which he is already familiar. Then the new article is placed alongside an old one on the same subject and they are compared, line by line, as to similarities and dissimilarities. If two chemists start with the same materials and wind up with the same products, it can be supposed that both used the same process, but this is only an assumption which must be checked by looking up unfamiliar words in the dictionary.

Three chemists scanning the same article may be looking for three entirely different types of information: one might need a date or a melting point, the other identification of the process, and the third sufficient information to enable him to repeat the experiment. The first objective would require little effort; the third, a great deal. Whether he wants little or much, the chemist should avoid hasty conclusions and make absolutely sure of the accuracy of what he does record. As a minimum, the chemist should be able to find out enough about an article to decide whether it is desirable to order a translation.

GENERAL APPLICABILITY OF THE DIRECT APPROACH

The method here outlined is generally applicable, since it is independent of the words of the language in which an article is printed. The examination of an article in an unfamiliar language is a slow process, but the chemist, remembering words learned with the first article, would find the second one easier. As he scans article after article, he builds up a vocabulary which enables him to go faster. In languages in which only few articles appear, speed is not necessary. He gains facility if, as, and when needed.

As foreign chemical articles are concentrated in German, French, and Russian, the chemist must acquire facility in these languages. Most chemists have some facility in French and German, but there are few who would not do well to add to it. Achieving facility in Russian is more difficult, but it is possible by learning the alphabet and applying the methods here described. For any one limited field, only a small vocabulary is required. As an example, the late John R. Sampey, who picked up his Russian from six articles by James W. Perry in the *Journal of Chemical Education*,[3] abstracted several hundred articles on chemotherapy for *Chemical Abstracts*.

People are prone to read words and try to remember them as words, sometimes without being conscious of their meanings. A chemist should make an effort to look through the words, whether they be English or Polish, and see the facts behind them. It should be like looking at the action on the stage without being conscious of the opera glass. The emphasis in this book is on the facts as the realities, independent of the words that describe them. As a chemist reads the description of a preparation, he should imagine himself weighing out the chemicals and performing the indicated operations.

Words such as "graduate" or "condensor" mean nothing to anyone who has not handled them. The meaning of a word is only what the reader's past experience enables him to understand when he reads the word. Try to describe a sunset to a blind man, or the song of a mockingbird to one who cannot hear!

A beginner in getting information out of foreign articles looks up each word in the dictionary, translates it, and then thinks of what it stands for. The chemist is urged to think in terms of standard laboratory procedures, and to connect the printed word directly with the object itself. This technique, difficult at first, becomes easier with practice.

[3] James W. Perry, "Chemical Russian, Self-Taught," *Journal of Chemical Education* 21: 393–98 (1944); 23: 22–27, 116–22, 248 (1946); 24: 28–45, 79–93 (1947).

2

Languages Using the Latin Alphabet

DEMONSTRATION OF THE DIRECT APPROACH

To demonstrate the general applicability of the direct approach, a chemical article in each of a dozen languages is examined. All of these languages are printed in the Latin alphabet. Except for Russian and Japanese, these are the only languages in which there are considerable numbers of chemical articles.

In illustrating the *principle* of the direct approach, a sample text in a language unrelated to English was purposely chosen. The language is Polish, which will force the chemist to rely on his knowledge of the subject, on the universal language of chemistry, and on the dictionary. If he can get results with an article in Polish, he should do equally well with other languages unrelated to English, and still better with languages related to English that contain cognates.

The reason for chosing Polish is its close resemblance to Russian. The similarity of Polish and Russian will help breach the barrier between the English-speaking chemist and a Russian text.

SAMPLE POLISH TEXT ANALYZED

The article chosen here for this demonstration is on the synthesis of L-adrenaline by S. Biniecki and S. Emilian (*Acta Polon. Pharm.* 20(3): 245–46, 1963; *CA* 62: 466). It will be noted that in twelve printed lines there are ten recognizable chemical names and formulae and eight international words. With

11

the aid of these, without knowing any Polish, the chemist should be able to identify the process as condensation of chloroacetic acid with pyrocatechol: $ClCH_2COOH + C_6H_4(OH)_2 \longrightarrow ClCH_2COC_6H_3(OH)_2$. The chemist knows that this is usually effected by heating equal weights of chloroacetic acid, pyrocatechol, and phosphorus oxychloride, protected from oxidation by SO_2, on a steam bath for forty-five minutes. The product is taken up in hot water containing traces of hydrochloric acid and sodium bisulfite. The solution is filtered hot and cooled to zero degrees. Crystals of 4-chloroacetylpyrocatechol, melting point 173°, separate out.

Below is the section of the Polish article:

Synteza L-adrenaliny

Otrzymywanie L-adrenaliny należy do preparatywnie trudnych. Liczne prace (1–5), dotyczące otrzymywania tego związku zbyt ogólnikowo podają warunki syntezy.

Część Doświadczalna

1. 4-Chloroacetylopirokatechina. Syntezę tego związku wykonano wzorując się w zasadzie na przepisie Hobermana (1).

Mieszaninę 10 g kwasu monochlorooctowego, 10 g pirokatechiny oraz 10 g świeżo destylowanego $POCl_3$ ogrzewano na wrzącej łaźni wodnej pod chłodnicą zwrotną (zabezpieczoną rurką z $CaCl_2$) w ciągu 1 godz., przepuszczając osuszony SO_2. Po oziębieniu wydzielił się osad surowej chloroacetylopirokatechiny, którą przekrystalizowano z 50 ml wody (z dodatkiem 2 g wodorosiarczynu sodowego) oraz węgla aktywnego.

The chemist writes down the recognizable words and deductions he makes:

... L-adrenaliny ... preparatywnie ... syntezy.
1. 4-Chloroacetylopirokatechina. Syntezę ... przepisie Hobermana (1).
... 10 g kwasu monochlorooctowego [this must mean "acid chloroacetic"] 10 g pirokatechiny ... 10 g ... destylowanego [distilled] $POCl_3$, ... łaźni wodnej [this must mean "heated on a water bath"] ... (... $CaCl_2$) [this $CaCl_2$ must have been used in a tube to help protect against the entrance of moisture from the air] ... 1 godz., [probably "1 hour"] przepuszczając osuszony SO_2 [SO_2 protects against oxidation, thus: while passing (in) dried SO_2]. ... chloroacetylopirokatechiny, ... przekrystalizowano ... 50 ml wody [boiling water would be used] (... [containing] 2 g wodorosiarczynu sodowego [sodium bisulfite, from the dictionary: wodor-hydrogen, siarka-sulfur, sod-sodium]) ... węgla [carbon, from dictionary] aktywnego [activated charcoal is commonly used to decolorize the solution].

As this is printed in the Latin alphabet, the chemist is helped not only by formulae and numerals but by chemical and international words and by being able to deduce the meanings of three Polish words from the context. He uses a dictionary in places where recognizable words are insufficient. Remembering the meanings of words learned in this and in subsequent articles, the chemist would build up a vocabulary which would eventually enable him to scan a Polish article almost as rapidly as one in English. Of course, the researcher would not go through the article laboriously as has been done here. He would glance down the page, noting the recognizable words, and would write in his notebook: "4-ClCH$_2$COC$_6$H$_3$(OH)$_2$ — usual process — CaCl$_2$ tube to protect from moisture is necessary." This is almost exactly what he would have done if the article had been in English. He would have spotted the key words and written down some notes. The only difference is that the Polish article would take a little longer to analyze.

Another reason for the choice of Polish to show how the direct approach works is its close resemblance to Russian. In the opening lines of the Polish article we find the following recognizable words:

4-Chloroacetylopirokatechina. Syntezę ... przepisie ... 10 g ... mo-nochlorooctowego, 10 g pirokatechiny ... 10 g ... POCl$_3$.

In the transliterated Russian version we note the same recognizable words only slightly altered:

4-Khloroatsetilpirokatekhin. Sintez ... metodu ... 10 g ... monokhlorouk-susnoi kisloty, 10 g pirokatekhina ... 10 g ... POCl$_3$.

This indicates that simply by learning the alphabet, the chemist can apply the direct approach to Russian articles as well as to those in Latin type. Chapter 3 will be devoted to helping the chemist do just that.

LANGUAGES RELATED TO ENGLISH

The languages considered in this chapter belong to two major groups, Germanic and Latin. The percentage of articles in current chemical literature that are published in each of these languages are, where calculable, as follows:

Germanic	Percent	Latin	Percent
German	9.8	French	5.1
Dutch	0.2	Italian	1.9
Swedish	0.1	Spanish	0.6
Danish	?	Romanian	0.6

German and French are the dominant languages. A knowledge of these two is of much help in dealing with the others of their group. Both of them are so closely related to English that, in addition to chemical names and international words, many words are recognizable on account of their cognate relationship. The French *six* (pronounced *sees*) and the English "six" look the same and sound differently, while the German *Eis* and the English "ice" are spelled differently and sound alike. In spite of this, many words are easily understood by the English-speaking reader.

English, historically of Germanic origin, has adopted so many words derived from Latin that it is closely related to French and the other Latin languages. Thus we have:

German	English		French
hundert	hundred	century	cent
zehn	ten	decimal	dix
Mann	man	homicide	homme
Vater	father	paternal	père
Haus	house	mansion	maison
Eis	ice	glacier	glace

In chemical articles, the proportion of French words recognizable from their similarity to English is more than double that of the German. One must be cautious, however. What looks like an English word may mean something entirely different. In French, *parent* may be any relative, and *hôtel de ville* is not a place to get a night's lodging. In German, *man* does not mean "man," but is an indefinite pronoun. *Man sagt* means "they say." The German word for "man" is *Mann*.

The languages of the Latin and Germanic groups are the languages of Western Europe, spoken by peoples who have long been in close commercial and cultural contact. All of them are of the same linguistic ancestry and still have much in common.

Since the time of Galileo (1564–1642) interest in the natural sciences has spread northward from Italy to England, France, and Germany, and results of investigation have been published in the languages of these countries.

English, German, French, and Italian, spoken by Priestly, Dalton, Liebig, Lavoisier, and Avogadro, became the dominant languages. As has been mentioned in Chapter 1, most chemists need little help with these languages. The object of this chapter is to demonstrate the applicability of the direct approach to articles of different character and in different languages. There are many young researchers who need the help offered in this section and it is believed that even the more experienced chemists will get something of value from it. It is interesting to trace the relationships of the languages of a group to each other. For this purpose several languages of a group are combined in a single glossary. (See Appendix A.)

THE LATIN GROUP

Italian, spoken now in the country of ancient Romans, is closest to Latin. French, Spanish, and Portuguese show strong Western influence and of all languages offer the least difficulty to English-speaking chemists. Romanian, which has been in contact with Eastern languages, is further removed from English than is Italian. One advantage with all of these languages is that the sentence structure is similar to that in English.

ITALIAN. There are many chemical articles in Italian; about forty percent as many as in French. They cause little difficulty to those who are acquainted with French or Latin.

The experimental section of an Italian article on the preparation of 4-thiocyanobiphenyl by Nerina Cagnoli (*Boll. Chim. Farm.* 97: 660–62 [1958]; *CA 53*: 8058) is chosen for examination:

Parte sperimentale

4-solfocian-bifenile (I)

Si ottiene diazotando 10 g di 4-aminobifenile in soluzione di acido formico e acido cloridrico, con 4,2 g di nitrito di sodio e decomponendo il diazo in presenza di solfocianuro rameoso.

Cristalli bianchi da alcool diluito, di forte odore aromatico; p.f. 84°.

4-4'-disolfocianbifenile (II)

Si ottiene come I impiegando 10 g di benzidina e 7,6 g di nitrito di sodio. Aghi gialli da alcool diluito; p.f. 129°.

La nitrazione di I e II fu eseguita sciogliendo i composti a 0° in acido solforico conc. e aggiungendo lentamente a −5° il nitrato di etile.

Dinitro-4-solfocianbifenile (III)

Ciuffi di aghi gialli da alcool etilico; p.f. 186–187°.

Dinitro-4-4'-disolfocianbifenile (IV)

Aghi bianchi in ciuffi; p.f. 174°.

2-amino-6-fenil-benzotiazolo (V)

10 g di 4-aminobifenile si sciolgono in acido acetico glaciale, si aggiungono 12 g di solfocianuro di potassio in fine polvere e in tale soluzione si fa sgocciolare lentamente la soluzione acetica di 3,5 ml di bromo. Si getta la massa di reazione in acqua e si raccoglie il precipitato; esso, anche dopo cristallizzasione, non ha un punto di fusione ben definito: 260–270°, poichè è una miscela di derivato tiazolico e di 4-amino-3-rodan-bifenile. Si bolle allora in acido cloridrico 1:1, si neutralizza la soluzione e si cristallizza da acido acetico.

Cristalli bianchi; p.f. > 310°.

Glancing at the article, the organic chemist spots the word *diazotando* which gives the whole thing away. From his experience the chemist would know what chemicals are to be used, their relative amounts, and the reaction conditions. There are a number of articles in chemical literature describing the preparation of aromatic thiocyanates by diazotizing an aromatic amine and coupling with copper thiocyanate. It might well be that all the chemist would want to know from this article is that this particular aromatic thiocyanate can be prepared by the standard method. However, he might be interested in the details of the preparation, which he can learn by going through the article, taking recognizable words, and using the dictionary for others, the meanings of which are in parentheses:

Parte sperimentale

4-solfocian-bifenile (I)
. . . diazotando 10 g . . . 4-aminobifenile in soluzione . . . acido formico . . . acido cloridrico, . . . 4,2 g . . . nitrito . . . sodio . . . decomponendo . . . diazo in presenza . . . solfocianuro rameoso [copper].
   Cristalli bianchi [white] . . . alcool diluito, . . . odore aromatico; p.f. [point of fusion] 84°.

The rest of the article concerning the diazotization of benzidene is left as an exercise.

SPANISH.   Spanish and Portuguese, spoken not only in Europe but also in South America, Central America, and a considerable portion of North America, may be expected to become increasingly important as languages of chemistry.
   Part of an article by E. F. Recondo and H. Rinderknecht (*Anales Asoc. Quim. Arg.* 47: 312–17 [1959]; *CA 55*: 15356) on the preparation of a pentose derivative is presented as an example:

Parte Experimental

   1-0-acetil-2,3,5-0-tribenzoil-β-D-ribofuranósido. — 75 gramos de D-ribosa se disolvieron en 1600 ml de metanol anhidro por agitación mecánica. Se agregaron entonces 72 ml de una solución de metanol que contenía 6,6 g de ácido clorhídrico y se continuó la agitación hasta que la solución dió reacción prácticamente negativa para grupos reductores (a) (1,¼ horas). Inmediatamente se detuvo la reacción por el agregado de 150 ml de piridina anhidra (b) y se concentró la solución bajo presión reducida con la temperatura del baño debajo de 50° (c) hasta eliminación de los solventes.
   1-0-metil-2,3,5-0-tribenzoil-D-ribofuranosido. — Se añadieron 150 ml más de piridina y la solución se concentró nuevamente a sequedad (d), el líquido amarillo remanente se disolvió en 400 ml de cloroformo y 880 ml de piridina anhidra. La solución se enfrió hasta alrededor de 10–15°C (temperatura interna) y se agregaron lentamente con agitación 292 ml de cloruro de benzoilo (e) de manera tal de conservar la temperatura interna debajo de los 30°. La mezcla heterogénea se dejó a 5° durante la noche.

Al día siguiente se vertió en agua helada con agitación; se separó la fase orgánica y se extrajo la fase acuosa con dos porciones de 200 ml de cloroformo. La fase orgánica y los extractos clorofórmicos se lavaron cuidadosamente con agua, se secaron con sulfato de sodio y se filtraron. La solución pardo amarillenta fué evaporada bajo presión reducida hasta eliminación de los solventes (temperatura del baño debajo de 70°).

Disregarding configurational differences, a pentose may be written:

$$CHO-CHOH-CHOH-CHOH-CH_2OH$$

This is the aldehyde form which is in tautomeric equilibrium with the furanose:

$$\overset{\displaystyle \lceil\!\!-\!\!-\!\!-\!\!-\!O-\!\!-\!\!-\!\!-\!\rceil}{HO-CH-CHOH-CHOH-CH-CH_2OH}$$

This is called furanose since the ring contains four carbon atoms and one oxygen as in furan. Treatment of D-ribose with methanol and an acid converts it to the methyl furanoside:

$$\overset{\displaystyle \lceil\!\!-\!\!-\!\!-\!\!-\!O-\!\!-\!\!-\!\!-\!\rceil}{MeO-CH-CHOH-CHOH-CH-CH_2OH}$$

The three remaining hydroxyl groups are easily benzoylated. These reactions as applied to D-ribose are easily followed by taking recognizable words with help from the dictionary:

Parte Experimental

1-0-acetil-2,3,5-0-tribenzoil-β-D-ribofuranósido. — 75 gramos de D-ribosa . . . disolvieron en 1600 ml de metanol anhidro . . . agitación mecánica. . . . 72 ml . . . solución de metanol . . . contenía 6,6 g de ácido clorhídrico . . . continuó la agitación hasta [until] . . . solución . . . reacción prácticamente negativa para grupos reductores (a) (1,¼ horas) [i.e. until no reaction with Fehlings solution — approx. one and a quarter hours]. Inmediatamente . . . reacción . . . 150 ml de piridina anhidra (b) . . . concentró la solución bajo [under] presión reducida . . . temperatura del baño [bath] debajo de 50° (c) hasta eliminación . . . solventes.

The rest is left as an exercise.

ROMANIAN. There are more chemical articles from Romania than from several much larger and better known countries. The Latin language was introduced by the Romans about 100 A.D. and has absorbed much from the languages of nations that overran the country. The sentence structure is similar to that of English, so that Romanian chemical articles are not difficult to scan.

The article chosen for examination concerns the reactions of α-phenoxypropionyl chloride with glycine and related compounds. Two of the six parts of the experimental section are given below but only the first is examined. The remainder may be used as an exercise.

Here are the reactions for the two parts of the experimental section:

I. $MeCH(OPh)COCl + H_2NCH_2COONa$

$$\longrightarrow MeCH(OPh)CONHCH_2COONa$$

II. $MeCH(OPh)COCl + H_2NCH_2CN$

$$\longrightarrow MeCH(OPh)CONHCH_2CN$$

Following is the original Romanian article by V. Voinescu and A. Balough (*Rev. Chim. (Bucharest)* 15(11): 677–78 [1964]; *CA* 62: 11718b).

Partea experimentală

I. *Acidul α-fenoxipropionilaminoacetic.* a) La soluția formată prin dizolvarea a 10 g bicarbonat de sodiu în 120 ml apă se adaugă 2,25 g (0,03 M) glicină. Sub răcire și agitare se picură 6,26 g (0,031 M) clorură de acid α-fenoxipropionic în 10 ml acetonă. Se continuă agitarea 1½ ore la 15–18°C. Soluția se spală prin agitare de două ori cu cîte 20 ml metilizobutilcetonă. Se acidulează cu acid clorhidric la pH = 2, cînd apare un ulei. Se extrage cu 40 ml acetat de etil, se usucă pe sulfat de sodiu anhidru și se distilă in vid. Reziduul se reia cu apă și se filtrează. Se obțin 5,2 g (77,7%) produs cu p.t. = 105°C.

b) 2,37 g (0,01 M) din (V) se refluxează cu 0,4 g (0,01 M) hidroxid de sodiu în 30 ml metanol, timp de 4 ore. După distilarea metanolului în vid, reziduul se dizolvă în apă și se filtrează cu cărbune. Filtratul se acidulează cu acid clorhidric la pH = 2, se filtrează precipitatul depus și se spală pe filtru cu apă.

Se obține 1,8 g (81%) produs; p.t. = 106,5°C după recristalizare din eter de petrol.

II. *α-fenoxipropionilaminoacetonitril.* 36,8 g (0,2 M) de clorură a acidului α-fenoxipropionic și 22 g (0,22 M) clorhidrat de aminoacetonitril se refluxează în 300 ml benzen anhidru pînă la încetarea degajării de HCl. După distilarea benzenului, produsul de reacție se toarnă peste soluția apoasă de bicarbonat de sodiu, se filtrează și se spală cu apă pe filtru. Se usucă.

Se obțin 36,8 g (90%) produs brut; p.t. după purificare din benzen-eter de petrol a fost 84°C.

Taking the recognizable words with some help from the dictionary we have the following:

Partea experimentală

I. *Acidul α-fenoxipropionilaminoacetic.* a) ... soluția formată ... dizolvarea ... 10 g bicarbonat de sodiu în 120 ml apă [water] ... adaugă [add] 2,25 g (0,03 M) glicină. ... răcire [cool] ... agitare ... 6,26 g (0,031 M) clorură de acid α-fenoxipropionic în 10 ml acetonă. . . . continuă agitarea 1½ ore [hours] . . . 15–18°C. Soluția . . . spală [wash] . . . agitare . . . două [half] ori cu [with] . . . 20 ml metilizobutilcetonă. . . . acidulează cu acid clorhidric . . . pH = 2, . . . extrage cu 40 ml acetat de etil, . . . usucă [dry] . . . sulfat de sodiu anhidru . . . distilă în vid. Reziduul . . . reia [wash] cu apă . . . filtrează. . . . obțin 5,2 g (77,7%) produs cu p.t. = 105°C.

b) 2,37 g (0,01 M) . . . refluxează cu 0,4 g (0,01 M) hidroxid de sodiu în 30 ml metanol, . . . 4 ore. . . . distilarea metanolului în vid, reziduul . . . dizolvă în apă . . . filtrează cu cărbune. Filtratul . . . acidulează cu acid clorhidric . . . pH = 2, . . . filtrează precipitatul . . . spală . . . filtru cu apă. . . . obține 1,8 g (81%) produs; p.t. = 106,5°C. . . . recristalizare . . . eter de petrol.

FRENCH. The experiments described above show that by following recognizable words a chemist who is well informed on the matter at hand can get much chemical information out of chemical articles written in any one of these Latin languages. With moderate help from dictionaries he can even make fair translations. In these languages a great deal of help is derived from cognate words, the meanings of which are evident from their resemblance to the English.

French is next to German and Russian in its importance as a language of chemistry. On account of the abundance of cognate words and of the similarity of sentence structure to that of English, French gives little trouble to the chemist. English is bad enough in the scant relationship of spelling to pronunciation. There seems to be no good reason that "bow" as in "bow-and-arrow" should be pronounced like "beau," while "bow" as part of a ship is pronounced "*bough.*" French is much worse, though with some practice it is usually possible to look at the printed word and come close to the pronunciation. Oral instruction is necessary for this. Fortunately the chemist does not have to spell the words.

The French negative is also confusing. We say: It is not true. For emphasis we can say: It is not a bit true. The French cannot say: Ce n'est vrai; but must say: Ce n'est pas vrai. *Pas* means *step* but takes the place of our *bit*. The French negative, *ne*, can be used only in front of a verb. We say: Not true. As there is no verb, *ne* cannot be used. So they say: *Pas vrai.*

The desirable facility in reading French chemical articles is acquired by extensive practice. It is recommended that the chemist read article after article without bothering about the meaning of strange words. He may not do very well on the first but should do better on the hundredth. The beginner sees a French word and goes by way of the dictionary to the English word and then to the object. When reading a French article the mind should be, as far as possible, devoid of English words.

The chemist's attention is called to a delightful article by J. Liebig titled "Ueber das Substitutionsgesetz und die Theorie der Typen" [About the Substitution Law and the Type Theory], *Ann. Chem.* 33–34: 308–10 (1840). This article is a take off on the substitution theory which at that time was very controversial.

According to the dualistic theory propounded by Berzelius, the elements were divided into two classes, electro-positive and electro-negative. The combination of manganese and oxygen, MnO, would still be electro-positive. As sulphur was electro-negative, its oxide would still be electro-negative. The combination of the two, manganese sulphate, was written $MnO \cdot SO_3$. It seemed preposterous that an electro-negative element such as chlorine could be substituted for electro-positive hydrogen. In a letter replying to Liebig's article, S. C. H. Windler goes still farther and substitutes chlorine for one after another of the elements in manganese acetate:

Monsieur,   Paris le 1er mars 1840.
Je m'empresse de vous communiquer un des faits les plus éclatants de la chimie organique. J'ai vérifié la théorie des substitutions d'une manière extrêmement remarquable et parfaitement inattendue. C'est seulement dès à présent qu'on pourra apprécier la haute valeur de cette théorie ingénieuse et qu'on pourra entrevoir les découvertes immenses, qu'elle nous promet de réaliser. La découverte de l'acide chloracétique et la constance des types dans les composés chlorés, dérivés de l'éther et du chlorure d'éthyle m'ont conduit aux expériences que je vais maintenant décrire. J'ai fait passer un courant de chlore à travers une dissolution d'acétate de manganèse, sous l'influence directe de la lumière solaire. Après 24 heures j'ai trouvé dans le liquide une superbe cristallisation d'un sel jaune violacé. La dissolution ne contenait que le même sel et de l'acide hydrochlorique. J'ai analysé ce sel: c'était du chloracétate de protoxyde de manganèse. Jusqu'ici rien d'extraordinaire, simple substitution de l'hydrogène de l'acide acétique par un nombre d'équivalents égaux de chlore, déjà connue par les belles recherches sur l'acide chloracétique. Ce sel chauffé à 110° dans un courant de chlore sec fut converti avec dégagement de gaz d'oxygène en un nouveau composé jaune d'or, dont l'analyse conduisait pour sa composition à la formule $MnCl_2 + C_4Cl_6O_3$. Il y avait donc substitution de l'oxygène de la base par du chlore, ce qu'on a observé dans une foule de circonstances. La nouvelle matière se dissolvait dans du chloral bien pur à l'aide de la chaleur, et je me servis de ce liquide, inaltérable par le chlore, pour pouvoir continuer le traitement par cet agent. J'y fis passer du chlore sec, pendant 4 jours, en tenant le liquide toujours très près de son point d'ébullition. Durant ce temps il se déposa constamment une matière blanche, qui dans un examen attentif, fut reconnu pour du protochlorure de manganèse. Je fis refroidir le liquide quelque temps après, où il n'y avait plus de précipité et j'obtins un troisième corps en petites aiguilles, soyeux jaune verdâtre. C'était $C_4Cl_{10}O_3$, ou en d'autres termes c'était l'acétate de manganèse dans lequel tout l'hydrogène et l'oxyde de manganèse étaient remplacés par du chlore. Sa formule devra être écrite $Cl_2Cl_2 + C_4Cl_6O_3$. Il y avait donc 6 atomes de chlore dans l'acide, les quatre autres atomes représentaient l'oxyde de manganèse. Comme l'hydrogène, le manganèse et l'oxygène peuvent être remplacés par le chlore, on ne verra dans cette substitution rien de surprenant.
Mais ce n'était pas encore la fin de cette série remarquable de substitutions. En faisant agir de nouveau le chlore sur une dissolution de cette matière dans l'eau, il y avait dégagement d'acide carbonique et en refroidissant le liquide à $+2°$ il se déposa une masse jaunâtre formée de petites paillettes,

ressemblant extrêmement à l'hydrate de chlore. Aussi ne contenait elle que du chlore et de l'eau. Mais en prenant la densité de sa vapeur j'ai trouvé qu'elle était formée de 24 atomes de chlore et de 1 atome d'eau. Voilà donc la substitution la plus parfaite de tous les éléments de l'acétate de manganèse. La formule de la matière devra être exprimée par $Cl_2Cl_2 + Cl_8Cl_6Cl_6 + aq$. Quoique je sache, que dans l'action décolorante du chlore il y a remplacement de l'hydrogène par le chlore et que les étoffes, qu'on blanchit maintenant en Angleterre d'après les lois de substitutions, conservent leur type, je crois néanmoins que la substitution du carbone par le chlore atome pour atome, est une découverte qui m'appartient. — Veuillez bien en prendre note dans votre journal et recevez etc.   S. C. H. Windler.

COMPARISON OF THE LATIN LANGUAGES.   For comparison of the five languages with each other and with English, translations of a section of Sabatier's *Catalysis in Organic Chemistry* are presented. Each of these is in the native tongue of the chemist making the translation. It is recommended that, without reading the English, the reader attempt a translation of each of the versions and then compare them with the English.

The close similarity of Portuguese to Spanish is seen in a comparison of the two versions of the Sabatier section. On account of this and of the fewness of chemical articles in Portuguese it does not seem necessary to scan an original article in it.

A comparison of the five Latin-language versions shows that the chemical and international words are nearly the same in all of them, with the exception that the number of cognates is low in the Romanian. The numbers of these recognizable words are as follows:

Recognizable Words	French	Italian	Spanish	Romanian	Portuguese
Chemical	7	7	7	7	7
International	13	13	13	12	13
Cognate	21	21	22	12	22

Comparison of the figures for the cognate words substantiates statements made above. The chemical and international terms show almost no variation.

The Catalytic Hydrogenation of Acetylene

French — Original by Paul Sabatier

L'acétylène se combine à froid avec l'hydrogène au contact du noir de platine, et fournit successivement de l'éthylène, puis de l'éthane.

En présence d'hydrogène en excès, l'acétylène est totalement transformé en éthane pur, sans aucune production accessoire.

A 180°, la même réaction a lieu plus vite, mais il y a formation d'une certaine dose d'hydrocarbures supérieurs liquides. En augmentant la proportion d'acétylène dans le mélange, l'éthylène domine dans les produits, mais il subsiste toujours de l'éthane, même à côté d'acétylène non transformé.

Si la dose de ce dernier devient assez grande, le noir de platine étant à 180°, on observe une certaine destruction charbonneuse du gaz, qui finit par fournir une incandescence analogue à celle que donne le nickel.

La mousse de platine n'exerce à froid aucune action, et n'est active qu'au-dessus de 180° pour l'hydrogénation de l'acétylène.

Italian — Translated by L. Mazza

L'acetilene si combina a freddo con l'idrogeno a contatto del nero di platino e fornisce successivamente dell'etilene, poi dell'etano.

In presenza di idrogeno in eccesso l'acetilene è totalmente trasformato in etano puro, senza alcun prodotto accessorio.

A 180° la stessa reazione ha luogo piu velocemente, ma si ha formazione di una certa quantita di idrocarburi superiori liquidi. Aumentando la proporzione di acetilene nel miscuglio, l'etilene domina nei prodotti, ma rimane sempre dell'etano insieme ad acetilene non trasformato.

Se la quantità di quest'ultimo diviene abbastanza grande, essendo il nero di platino a 180°, si osserva una certa distruzione carboniosa del gas, che finisce per dare una incandescenza analoga a quella che da il nichel.

La spugna di platino a freddo non esercita alcuna azione e non è attiva che al di sopra di 180° per l'idrogenazione dell'acetilene.

Spanish — Translated by M. Baron

Por contacto con negro de platino el acetileno se combina en frio con hidrógeno para dar en forma sucesiva etileno y etano.

En presencia de un exceso de hidrógeno el acetileno se transforma totalmente en etano puro, sin que aparezca ningún producto de reacción secundario.

La misma reacción se produce a 180° más rápidamente pero con formacion de una cierta cantidad de hidrocarburos superiores líquidos. Incrementando la proporción de acetileno en la mezcla hay un predominio de etileno en los productos, pero aun así se encuentra siempre algo de etano, aun en presencia de acetileno no transformado.

Con el negro de platino a 180° y si la proporción de acetileno se incrementa lo suficiente, se observa una destrucción carbonosa parcial del gas, que llega a dar una incandescencia semejante a la que produce el niquel.

En frío, el musgo de platino no tiene influencia alguna, resultando activo para la hidrogenación del acetileno solamente por encima de 180°.

Romanian — Translated by F. Schneidman

Acetilena prin combinație cu hidrogenul la rece in prezența negrului de platin dă întîi etilena și apoi etan.

In prezența unui exces de hidrogen, acetilena este complect transformată in etan pur, fără reacții secundare.

La 180°, aceiași reactie are loc mai rapid, dar se formează și anumite cantitati de hidrocarburi in alte lichide.

Prin marirea proportiei de acetilena in amestec, etilena devine principalul produs, dar ceva etan se formează, totdeauna chiar prin neschimbarea acetilenei rămase.

Daca proportia de acetilenă devine destul de mare, cu negru de platin la 180°, o anumita cantitate de fum a gazului descompus este observată și aceasta se termină cu incandescența, la fel ca in cazul nichelului.

Platinul poros nu este activ la rece și nu are nici un efect la hidrogenarea acetilenei decit peste 180°.

Portuguese — Translated by W. B. Mors

O acetileno se combina com hidrogênio a frio na presença de negro de platina, dando primeiro etileno e em seguida etano.

Na presença de um excesso de hidrogênio, acetileno é inteiramente transformado em etano puro, sem quaisquer reações secundárias.

A 180° a mesma reação se dá mais ràpidamente, mas verificase a formação de certa quantidade de hidrocarbonetos superiores, líquidos. Aumentando-se a proporção de acetileno na mistura, o etileno torna-se o produto principal, mas certa quantidade de etano sempre se forma, não obstante a presença residual de acetileno não transformado.

Se, com o negro de platina a 180°, a proporção de acetileno é tornada suficientemente elevada, observa-se um certo gráu de decomposição fuliginosa do gás, terminando em incandescência, como no caso do níquel.

Esponja de platina não mostra atividade a frio e não efetua a hidrogenação do acetileno a não ser acima de 180°.

English — Translated by E. E. Reid

Acetylene combines with hydrogen in the cold in the presence of platinum black, giving first ethylene and then ethane.

In presence of an excess of hydrogen, acetylene is entirely transformed into pure ethane without any side reactions.

At 180° the same reaction takes place more rapidly but there is the formation of a certain amount of higher liquid hydrocarbons. By augmenting the proportion of acetylene in the mixture, ethylene becomes the main product but some ethane is always formed even though unchanged acetylene remains.

If the proportion of the acetylene becomes great enough, with the platinum black at 180°, a certain amount of smoky decomposition of the gas is observed and this ends with incandescence, as is the case with nickel.

Platinum sponge is not active in the cold and does not effect the hydrogenation of acetylene except above 180°.

THE GERMANIC GROUP

This group consists of five closely related languages: German, Dutch, Swedish, Norwegian, and Danish. Scheele, the codiscoverer of oxygen, and Berzelius, one of the greatest of the early chemists, spoke Swedish. The "Four Dutch Chemists" (Deimann, Troostwyk, Bondt, and Lauwerenburgh) and van't Hoff spoke Dutch, and Zeise, the father of organic sulfur chemistry, spoke Danish, but all of them wrote in German, which dominates the group.

Compared to the Latin group, the Germanic languages have about the same number of chemical words but fewer international and cognate terms. Figures for the various versions of the Sabatier section are given later.

The sentence structure in Dutch, Norwegian, Swedish, and Danish is very similar to that in English. Except that he will have to use the dictionary more freely, the chemist will find articles in these languages are almost as easy to scan as those in the Latin group.

Because of differences in sentence structure, German chemical articles offer considerably more difficulty. On this account and because of the high proportion of early chemical articles in German, more attention and space is given to German, in a separate section. Most chemists studied German in preparation for a career in chemistry but many lack facility in its use.

DUTCH.   Dutch chemists have contributed much to science but have published most of their work in other languages. Corresponding to the geographical position of the country in which it is spoken, Dutch is about halfway between German and English. The meanings of Dutch words which are not recognizable from their resemblance to English may usually be guessed from a German dictionary.

For a study, a portion of an article on the preparation of sodium diethyldithiocarbamate by G. J. van Kolmeschate and R. W. Stern (*Chem. Weekblad* 45: 733–34 [1949]; *CA* 44, 3903) follows:

Reactievergelijking:

$(C_2H_5)_2NH + CS_2 + NaOH \longrightarrow (C_2H_5)_2NCS_2Na + H_2O$

Nodige chemicaliën:
7.3 g diaethylamine, opgelost in 10 ml aethanol,
7.6 g zwavelkoolstof,
4   g natriumhydroxyde, opgelost in zo weinig
        mogelijk water: de oplossing wordt verdund
        met 10 ml aethanol.
Voorschrift: In een kolfje worden het diaethylamine en de zwavelkoolstof voorzichtig gemengd. Zo nodig wordt onder de waterstraal gekoeld. Soms scheidt zich bij deze reactie een vaste stof af, hetgeen echter voor het verdere verloop der reacties geen enkel nadeel heeft. Daarna voegt men de loogoplossing druppelsgewijs toe. Het reactiemengsel wordt op kamertemperatuur afgekoeld en dan uitgegoten in een overmaat aether. Er scheidt zich dan een witte kristalbrij af. Deze wordt afgezogen en met aether nagewassen.

Opbrengst bijna 80%. Eventueel kan men de kristallen zuiveren door ze in zo weinig mogelijk aethanol op te lossen en de oplossing in aether uit te gieten. Ook zonder deze zuivering verkrijgt men een praeparaat, dat ook uiterlijk beter is dan sommige praeparaten uit de handel. De aldus verkregen verbinding voldoet aan de eisen in Merck's *Prüfung auf Reinheit* 5. Aufl., S. 420 daaraan gesteld.

The details of the experiment are easily followed by taking the recognizable words with some help from the dictionary:

Reactievergelijking:

$$(C_2H_5)_2NH + CS_2 + NaOH \longrightarrow (C_2H_5)_2NCS_2Na + H_2O$$

... chemicaliën:

7.3 g diaethylamine, opgelost [dissolved] in 10 ml aethanol,

7.6 g zwavelkoolstof,

4 g natriumhydroxyde, opgelost in ... weinig mogelijk [least possible] water: ... oplossing ... verdund [diluted] ... 10 ml aethanol.

... In ... kolfje [German *kolbe* — flask] ... diaethylamine ... zwavelkoolstof ... gemengd [mixed]. ... onder ... waterstraal [stream of water] gekoeld. ... reactie ... reacties .... [The first reaction is: $2Et_2NH + CS_2 \longrightarrow$ $Et_2NCSSNH_2Et_2$. It is known that the diethylammonium diethyldithiocarbamate may separate out.] ... loogoplossing [alkali solution] druppelsgewijs ... reactiemengsel ... kamertemperatuur [room temperature] afgekoeld ... overmaat [excess] aether. ... witte kristalbrij ... aether nagewassen [washed]. Opbrengst ... 80%. Eventueel ... kristallen ... weinig mogelijk aethanol ... lossen ... oplossing in aether ... gieten [German *giessen* — pour]. ... praeparaat, ... beter ... praeparaten ...

THE SCANDINAVIAN LANGUAGES — *by Nis H. Skau*

The Scandinavian languages are so closely related that translations from one to another are scarcely necessary for any educated person who knows any of them with the exception of Icelandic. They belong to the Germanic family and it may be safely concluded that they had a common North Gothic ancestor close to modern Icelandic, which has changed little in a thousand years. Danish has been most affected by cultural influences from the south, marked particularly by the religious Reformation and later by the so-called Enlightenment. This is the origin of many German loan words entirely aside from the natural cognates. The ancient dialects survived longer in the isolated valleys of the Scandinavian peninsula, but by the time the idioms congealed into the present-day literary forms the cultural and political influences were predominant. Present-day Swedish is the language of the southern provinces influenced by their many-sided connections with Denmark and Germany. Norwegian *Bokmal* in a very real sense *is* Danish, the result of three hundred years of political union with Denmark which was followed by one hundred years of union with Sweden that had little effect on the language.

Structurally the Scandinavian languages resemble English, with their simplified grammar, nearly complete abandonment of case endings, and person-denoting endings of the verbs. But in contrast to English, they have continued the vowel changes in plural formation and in the tenses of the irregular verbs. They have two genders, the common gender which is a merger of the masculine and the feminine, and the neuter gender.

Entirely aside from the specific scientific terms, the overwhelming part of the vocabulary of the three languages is made up of cognates. The foreign reader of scientific literature who becomes acquainted with Danish-Norwegian may have a little more difficulty with Swedish and vice versa. That literary Norwegian and Danish are more closely related to each other than either of them is related to Swedish becomes evident through almost any word list selected at random. If sometimes the impression is obtained that Swedish and Norwegian have something in common that is not shared by Danish, this is almost entirely due to two factors. First, Swedish and Norwegian often have hard consonants where Danish has soft ones; *t* for *d*, *k* for *g*, *p* for *b*; secondly, Swedish and Norwegian use a nearly up-to-date, modernized spelling where Danish, like English, maintains an antiquated one. Because of a certain fluidity of Danish pronunciation, a really rational spelling reform in Danish would be next to impossible. Orthographically, all three use the *å* to designate a kind of open *o* sound similar to English *a* in "water." Until about 1930, Danish used the double *a*, *aa*, to designate this sound. Danish and Norwegian use the letters *æ* and *ø* for what corresponds to German *ä* and *ö*; similarly they use the *y* for what corresponds to the *ü*, although it is generally not recognized as such. Swedish follows the German custom writing *ä*, *ö*, and *ü*, sometimes *a*, *o*, and *u*.

Finnish is unrelated to the Scandinavian tongues and belongs to the Finno-Ugrian family of languages together with Estonian and Hungarian. During centuries of union with Sweden, however, as well as during well over a century of association with Russia, the administrative language of Finland was Swedish. Swedish is the native tongue of a little less than ten percent of the population. Now independent, the country is officially bilingual. All official documents, including patents, are published in both languages.[1]

SWEDISH. Below is the first part of the experimental section of an article by B. Sjöberg (*Svensk Kem. Tidskr.* 50: 150–54 [1938]; *CA* 33: 2106). It deals with the reactions of epichlorhydrin with hydrogen sulfide, which are so plainly stated that they need no comment:

Epiklorhydrin och svavelväte

Vid addition av svavelväte till epiklorhydrin i alkalisk lösning vid 0° erhölls tioklorhydrin:

$$\begin{array}{ccc} CH_2Cl & & CH_2Cl \\ | & & | \\ CH{\diagdown} & + H_2S = & CHOH \\ | \;\;{\diagup}O & & | \\ CH_2{\diagup} & & CH_2SH \end{array}$$

---

[1] See also section on Finno-Ugrian languages at the end of this chapter.

Tioklorhydrinen utgjorde en färglös, obetydligt luktande olja, lättlöslig i alkohol, eter, aceton och bensol, tämligen svårlöslig i vatten.

$$Kpt_{1,3} = 60°$$
$$n_D^{20} = 1,5257$$
$$d_{20} = 1,2981$$
$$MR_D = 29,91 \text{ för svavel } AR_D = 7,46$$

Vid 50° bildades vid additionen av svavelväte till epiklorhydrin oxytrimetylensulfid:

$$\begin{array}{c} CH_2 \\ | \\ CHOH \quad\rangle S \\ | \\ CH_2 \end{array}$$

Oxytrimetylensulfiden var en färglös, obetydligt luktande olja, lättlöslig i vatten, alkohol, eter, aceton och bensol.

Tioklorhydrinens konstitution fastställdes därefter. Tioklorhydrinen oxiderades kvantitativt med vätesuperoxid till kloroxypropansulfonsyra.

$$\begin{array}{c} CH_2Cl \\ | \\ CHOH \\ | \\ CH_2SO_3H \end{array}$$

Kloroxypropansulfonsurt natrium bildade vita kristaller, svårlösliga i alkohol, bensol, kloroform och ättiketer, tämligen lättlösliga i vatten. Smpt. 246–247°.

Reaktionshastigheten vid den alkaliska hydrolysen av kloroxypropansulfonsurt natrium bestämdes.

I analogi med bildningen av sulfid genom bimolekulär reaktion av merkaptid och halogenförening

$$RS + ClR' \longrightarrow RSR' + Cl^-$$

kan oxytrimetylensulfiden tänkas bildas ur tioklorhydrinens alkalisalt genom monomolekulär reaktion:

$$\begin{array}{cc} CH_2Cl & CH_2 \\ | & | \\ CHOH = CHOH \quad\rangle S + Cl^- \\ | & | \\ CH_2S^- & CH_2 \end{array}$$

Utförda försök visade, att tioklorhydrinens alkalisalt sönderföll i vattenlösning. Vid indunstning av vattenlösningen och extraktion med eter erhölls oxytrimetylensulfid.

The story in the article is as follows, taking the recognizable words:

Epiklorhydrin . . . svavelväte

. . . addition . . . svavelväte . . . epiklorhydrin . . . alkalisk lösning [solution] . . . 0° erhölls [is obtained] tioklorhydrin:

$$
\begin{array}{c}
CH_2Cl \\
| \\
CH\diagdown \\
| \quad\;\; O \quad + H_2S = \\
CH_2\diagup
\end{array}
\qquad
\begin{array}{c}
CH_2Cl \\
| \\
CHOH \\
| \\
CH_2SH
\end{array}
$$

Tioklorhydrinen . . . lättlöslig [very soluble] . . . alkohol, eter, aceton . . . bensol, . . . svårlöslig [sparingly soluble] . . . vatten [water].

$$Kpt_{1,3} = 60°$$
$$n_D^{20} = 1,5257$$
$$d_{20} = 1,2981$$
$$MR_D = 29,91 \text{ för svavel } AR_D = 7,46$$

. . . 50° . . . additionen . . . svavelväte . . . epiklorhydrin oxytrimetylen-sulfid:

$$
\begin{array}{c}
CH_2\diagdown \\
| \\
CHOH \quad\diagdown S \\
| \\
CH_2\diagup
\end{array}
$$

Oxytrimetylensulfiden . . . lättlöslig . . . vatten, alkohol, eter, aceton . . . bensol.

It is easy to follow the information given in the first part of the article. The Swedish word *svavelväte* must mean hydrogen sulfide, and *lättlöslig* suggests the German word *glattlöslich*, denoting easily soluble.

Additional physical data are given in the original. For the oxidation of thiochlorhydrin we have the following:

Tioklorhydrinens konstitution . . . Tioklorhydrinen oxiderades kvantitativt . . . vätesuperoxid . . . kloroxypropansulfonsyra.

$$
\begin{array}{c}
CH_2Cl \\
| \\
CHOH \\
| \\
CH_2SO_3H
\end{array}
$$

Kloroxypropansulfonsurt natrium . . . kristaller, svårlösliga . . . alkohol, bensol, kloroform . . . ättiketer [ethyl acetate], . . . lättlösliga . . . vatten. Smpt. 246–247°.

The rest of the article is left as an exercise.

DANISH.   The section chosen for study is from a Danish patent on obtaining tetraethylthiuramdisulfide in crystalline form. In a previous patent it was found that tetraethylthiuramdisulfide and two molecules of CCl_4 form an addition compound which crystallizes well. The removal of CCl_4 from this leaves good crystals of tetraethylthiuramdisulfide.

In this patent (Danske Medicinal- og Kemikalie-Kompagni Aktieselskab. Dan. 78,609, Dec. 20, 1954; *CA* 50: 1902) a similar addition compound with CBr_4 is described:

Udførelseseksempel

En blanding af 10 cm³ ætanol, 0,2 g tetraætyltiuramdisulfid og 0,7 g kulstoftetrabromid opvarmes under omrøring, indtil der er indtrådt fuldstændig opløsning. Ved køling af opløsningen til stuetemperatur udskilles der 0,5 g gule krystaller, der smelter ved 103–106°C. Ved bestemmelse af disses bromindhold viser de sig at bestå af tetraætyltiuramdisulfid, indeholdende 2 mol kulstoftetrabromid pr. mol af moderstoffet. Bromindholdet er i et forsøg fundet til 67,3%, medens det teoretisk skulle være 66,7%. Krystallerne er holdbare ved stuetemperatur og atmosfærisk tryk.

Når krystallerne opvarmes til 60°C under et vakuum svarende til 0,02 mm kviksølvsøjle, afgives kulstoftetrabromid, og der bliver porøse, hvide, gennemskinnelige tetraætyltiuramdisulfidkrystaller tilbage i kvantitativt udbytte. Partiklerne af tetraætyltiuramdisulfid har samme ydre størrelse og form som de kulstoftetrabromidholdige krystaller, hvoraf de er dannet.

Taking recognizable words with some help from the dictionary:

... blanding ... 10 cm³ ætanol, 0,2 g tetraætyltiuramdisulfid ... 0,7 g kulstoftetrabromid ... under omrøring [stirring], indtil ... fuldstændig [complete] opløsning [solution]. ... køling ... opløsningen ... stuetemperatur [room temperature] ... 0,5 g ... krystaller, smelter ... 103–106°C. ... bestemmelse [German *Bestimmung* — determination] ... bromindhold ... tetraætyltiuramdisulfid, ... 2 mol kulstoftetrabromid pr. mol ... moderstoffet. Bromindholdet ... fundet [found] ... 67,3%, ... teoretisk ... 66,7%. Krystallerne ... holdbare [German *haltbar* — steady] ... stuetemperatur ... atmosfærisk tryk [pressure].
... krystallerne ... 60°C under ... vakuum ... 0,02 mm ... afgives kulstoftetrabromid, ... bliver [German *bleiben* — remain] porøse, hvide [white], ... tetraætyltiuramdisulfidkrystaller ... kvantitativt ... Partiklerne ... tetraætyltiuramdisulfid ... samme ... form ... kulstoftetrabromidholdige [-containing] krystaller, ...

THE GERMAN LANGUAGE — by Otto K. Schmied

The ability to read and comprehend articles dealing with chemistry when these are written in German offers some difficulty to the student, not because of the scientific terms involved, for these are readily recognized because of their similarity to the English, but because of the structure of sentences in German. German has some of the characteristics of Latin in that it is still a highly inflected language which allows somewhat free placement of many of the elements composing a sentence.

The position of subjects and objects, adjectives and verbs does not follow the more or less rigid pattern of English — subject, verb, object. Long adjectival phrases frequently stand between the definite article and the noun, verbs break into two parts, and the whole verb or parts of it may be at the end of a clause or sentence, perhaps two or three lines away from the subject.

The purpose of the following lines is not to write a German grammar, but to offer some practical suggestions for the students to help them read and comprehend an article written in German without laboriously working out a literal translation.

Nouns: Nouns in German are always capitalized and thus readily recognizable. Some are identical in English and German, as *Hunger* — hunger, *Wanderer* — wanderer. The meaning of others is readily recognized once certain common variations are learned. The German *v* frequently becomes *f* in English, *Vater* — father, *voll* — full; German *t* becomes *d* in English, *Garten* — garden, *Bett* — bed, *halten* — hold; German *z* becomes *t* in English, *zu* — to, *Zahn* — tooth; German *d* becomes *th* in English, *denken* — to think; German *sch* becomes plain *s* or *sh* in English, *Schwindler* — swindler, *scharf* — sharp; German *k* becomes *c* in English, *kommen* — come, *kalt* — cold. German *au* may be *ou* in English, *Haus* — house. Some German words when pronounced will disclose their meaning because they sound much like the corresponding English; *Milch* — milk, *Hund* — hound, *Mutter* — mother, *hundert* — hundred.

The meaning of many long German words can be gotten when these are broken into the short words which, joined in German, form the long words. Thus, in the Sabatier section we have *Gasgemisch* for "mixture of gases" and *Kohlenstoffabscheidung* for "separation of carbon" (*Kohlenstoff* — carbon).

Definite Articles; Cases of Nouns: The English definite article *the* has three forms in German, *der, die, das*. These articles change and tell the way the nouns are used, as subjects or objects, or to indicate possession and the like. It is good to know and to recognize the subject and possessive form of the articles particularly. In English the possessive form of a noun is indicated either by an apostrophe followed by *s* or following *s* or by the use of the preposition "of." In German this is indicated by the change in the article from the subject form (nominative) to the possessive form (genitive), *der — des, die — der, das — des*. In general, when two nouns follow each other, the second one is likely to be in the possessive form. For example:

Die Verwendung des Eisenchlorids gestattet *die* industrielle *Erzeugung des* Kohlenstofftetrachlorid . . .
The use of . . . the production of . . .

Die Alkoholate *des Aluminiums* . . .
          . . . of aluminum

Adjectives: As in English, adjectives stand in front of the nouns they modify thus separating the article and the noun. In German long adjectival phrases are frequently used, where in English a relative clause would be used.

Eine *über nicht erwärmtes reduziertes Kobalt geführte* Mischung . . .
A mixture . . . which is passed over unheated (and) reduced cobalt

Relative Pronouns: Besides *welcher, welche, welches*, the forms of the definite articles are used as relatives, having the meanings of "who, whom, which, what, that," etc. The forms of *der, die, das*, etc. can be recognized as relatives by the position of the verb in these relative clauses, namely, at the end of the clause or sentence. Certain conjunctions such as *weil, dass, obgleich, wenn*, etc. have the same effect on the position of the verb.

> Die Temperatur wird einfach auf einem bis 450° anzeigenden Thermometer abgelesen, *das in der Rinne neben dem Rohre liegt* und sich verschieben lässt, so *dass die Gleichförmigkeit der Erhitzung kontrolliert werden kann.*
> ... which is lying beside the tube ...
> ... in order that the uniformity of the heating can be controlled ...

Verbs: Past action may be expressed either by a simple past tense or by a tense formed with an auxiliary verb and a past participle. There are three auxiliary verbs, *haben, sein,* and *werden.*

Simple past tenses of many verbs, called weak verbs, are formed by adding *te* or *ete* to the verb stem. Other verbs, called strong verbs, form the simple past tense by changing the vowel.

> Einige Chemiker *versuchten* ...
> 1825 *beobachtete Corenwinder* ...
> 1863 *gelang* [gelingen] Debus mit Hilfe ...

Compound past tenses in English use the verb *to have* in the active voice and *to be* in the passive voice. In German the following applies:

Transitive verbs (verbs with objects) use *haben.*

Intransitive verbs (change of place or condition) use *sein.*

Passive voice uses *werden. Werden* is also used for the future tense.

> *Active voice*: Die Jodwasserstoffsäure *hat* ... *angerichtet* ... [has caused]
> *Passive voice:* Hinter das ... Glührohr *wird* ... *eingeschaltet* ... [is inserted]

To form a future tense, the auxiliary *werden* is used with an infinitive. All infinitives in German end in *en*. Most participles have the prefix *ge-* and can be readily recognized, as *führen — geführt, legen — gelegt.* Verbs which begin with the prefix *be-, ent-, er-, ge-, ver-, zer-* do not have the prefix *ge-* in the participle. This is also true of many foreign words taken over and "germanized", as *reduzieren — reduziert, konzentrieren — konzentriert.* Participles of weak verbs end in *t* or *et,* those of strong verbs in *en.*

Prefixes: It is characteristic of German to give prefixes to verbs, sometimes for directional purposes, such as *hin-* and *her-,* or merely for emphasis, or again to give a new meaning to a verb. The prefixes *be-, ent-, er-, ge-, ver-, zer-* cannot be separated from the verb. Prepositions used as prefixes and other words are separated from the verb in the present and simple past tenses and are then placed at the end of the clause or sentence. The meaning of an entire sentence is not disclosed until the end of the clause or sentence is reached.

> Diese Wirkung *tritt* ... *ein* ... [result in]
> Doch *nimmt* die Aktivität schnell *ab* ... [decreases]
> ..... und *schreitet* bei der Siedentemperatur des Gemenges *fort* ... [goes on]

Conditional Sentences: The use of *if* to begin a conditional sentence in English has no effect on the order of words in the sentence. In German the conditional sentence beginning with *wenn* places the verb at the end of the clause or sentence. The use of *wenn* may be avoided (in Sabatier nearly always) by merely interchanging the positions of subject and verb. This is the usual order in questions, verb — subject, and punctuated with a question mark. Where this word order exists without a question mark, the sentence is conditional and, in English translation, it must begin with "if."

> Steigert man die Konzentration . . . . [If one increases the concentration . . . .]
> Verwendet man auf 300 Teile Benzol . . . . [If one uses 300 parts of . . . .]

Pronunciation: German words are pronounced as they are spelled. In addition it may be noted that the sounds of some letters are different from what they are in English.

The umlaut gives some trouble, but two dots over a vowel have the same effect as placing an *e* after it; ä is equivalent to *ae*. In English the *e*, which modifies the vowel, is separated from it by a consonant: ban, bane; van, vane.

*   *   *

As the majority are somewhat familiar with the language it does not seem necessary to scan the article below by J. Houben and H. Pohl (*Berichte* 40: 1303–307). It is on the preparation of dithioacetic acid from the methyl Grignard reagent and carbon disulphide:

$$CH_3MgI + CS_2 \longrightarrow CH_3CSSH$$

J. Houben und H. Pohl: Über Carbithiosäuren. II. Die
geschwefelte Essigsäure, CH₃.CS.SH.

[Aus dem I. chemischen Institut der Universität Berlin.]
(Eingegangen am 18.März 1907.)

Vor kurzem berichteten wir in dieser Zeitschrift[1]) über einige Synthesen schwefelhaltiger Säuren, einer bis dahin so gut wie unbekannten Klasse von Verbindungen, die Seitenstücke der Carbonsäuren vorstellen und mit dem Namen Carbithiosäuren belegt wurden. Sie sind frei von Sauerstoff und charakterisiert durch die Gruppe

.C(:S).SH,

die anscheinend stärkere saure Eigenschaften besitzt als der Carboxyl-rest. Diese Säuren sind außerordentlich unbeständig und besonders dem Luftsauerstoff gegenüber empfindlich. Sie gehen dabei unter Verlust des Sulfhydryl-Wasserstoffs in die Vertreter einer zweiten neuen Körperklasse, die Thioacyldisulfide, über

$$R.C \underset{S-S}{\overset{S\quad S}{<\quad>}} C.R,$$

deren sauerstoffhaltige Analoge bei der Elektrolyse fettsaurer Alkalien durch Zusammentritt zweier Anionen auftreten müßten, aber, wie bekannt, sehr unbeständig und daher erst auf anderem Wege erhalten worden sind[2]). Bei den Carbithiosäuren liegen die Verhältnisse umgekehrt. Die Thioacyldi-

---

[1] Diese Berichte *39*, 3219 [1906]. Siehe auch Inaugural-Dissertation von H. Pohl, Berlin 1907.

[2] Brodie, Pogg. Ann. *121*, 382. Siehe auch diese Berichte *29*, 1726 [1896] und Ann. d. Chem. *298*, 288.

sulfide sind durchweg beständige Verbindungen, während die entsprechenden Säuren, wenigstens in der aromatischen Reihe, erst durch ihre Salze und Abkömmlinge genauer charakterisiert werden konnten.

In der aliphatischen Reihe ist es uns aber nach manchen Fehlversuchen endlich gelungen, eine Reihe von Carbithiosäuren analysenrein zu gewinnen, zunächst auch das Seitenstück einer der bekanntesten Fettsäuren, die

Methyl-carbithiosäure oder geschwefelte
Essigsäure,
$CH_3.CS.SH.$

8.6 g Magnesiumband[3] wurden in bekannter Weise mittels 50 g Jodmethyl in 150 ccm absolutem Äther zu Methylmagnesiumjodid gelöst, die Lösung mit Eis-Kochsalz abgekühlt und 31.7 g Schwefelkohlenstoff langsam zugetröpft. Kommt die bald unter Sieden einsetzende Reaktion zum Stillstand, so hebt man den Kolben für kurze Zeit aus dem Kältegemisch. Die Reaktionsflüssigkeit färbt sich erst grün, dann dunkelgrün, dann dunkelrotbraun. Sie bleibt 24 Stunden in schmelzendem Eis stehen. Ein intensiver, die Verschlüsse durchdringender Geruch macht sich schon zu Beginn der Umsetzung bemerkbar und vermag, selbst in weiteren Entfernungen den Aufenthalt zu einem unerträglichen zu gestalten. Gut wirkende Abzüge und sonstige Vorsichtsmaßregeln versagen anscheinend dem entsetzlichen Geruch der Methylcarbithiosäure gegenüber vollständig, und das Arbeiten mit der Substanz stellt an die Geduld des Experimentators außergewöhnliche Anforderungen.

Zur Zersetzung wird die Reaktionsflüssigkeit, die sich manchmal in zwei Schichten scheidet, wieder mit Eis-Kochsalz abgekühlt und vorsichtig erst mit Eis, dann mit eisgekühlter Salzsäure versetzt, wobei Ströme von Schwefelwasserstoff entweichen. Die rotbraune ätherische Schicht wird abgehoben und mit den ätherischen Auszügen der wäßrigen vereinigt, sodann wieder mit verdünnter Sodalösung extrahiert und die so gewonnene Lösung von methylcarbithiosaurem Natrium abermals gründlich — etwa 5–6mal — ausgeäthert. Diese letzte Extraktion erwies sich zur Entfernung mercaptanartiger Verunreinigungen als nötig. Darauf wird mit eiskalter Salzsäure angesäuert und die frei gemachte Carbithiosäure in Äther aufgenommen, die Lösung über Natriumsulfat getrocknet und konzentriert. Es hinterbleibt ein rotgelbes Öl, das im Vakuum destilliert wird. Durch die Capillare leitet man an Stelle des Luftstromes trocknes Kohlendioxyd. Mehrmals wiederholte Destillation führt schließlich zu einer völlig konstant siedenden Fraktion von analysenreiner Methylcarbithiosäure. Sie wurde in einer Menge von 5.8 g erhalten, entsprechend einer Ausbeute von 17.9% der theoretisch möglichen. Ihr Siedepunkt liegt unter 15 mm Druck bei 37°.

Die Säure stellt ein r o t g e l b e s , intensiv widerlich und stechend riechendes Öl dar. Der Geruch erinnert gleichzeitig an Mercaptan, Allylsulfid und Essigsäure. Sie ist, obschon der Essigsäure vergleichbar, im Wasser so gut wie nicht löslich, wird jedoch von allen organischen Lösungsmitteln, wie Äther, Alkohol, Aceton, Eisessig, Chloroform, Petroläther, Benzol usw., reichlich aufgenommen, anscheinend in jedem Verhältnis. Geringe Spuren färben das Lösungsmittel deutlich gelb. Lackmus wird von ihr gerötet.

0.1556 g Sbst.: 0.1480 g CO_2, 0.0630 g H_2O. — 0.1389 g Sbst.: 0.7050 g $BaSO_4$ (Zersetzung im Rohr nach C a r i u s).

---

[3] Wir benutzen jetzt mit Vorteil sehr reaktive Magnesiumspäne aus der Aluminium- und Magnesiumfabrik H e m e l i n g e n bei Bremen.

$C_2H_4S_2$. Ber. C 26.04, H 4.37, S 69.59.
Gef. » 25.98, » 4.53, » 69.68.

Da die Substanz ganz frei von Sauerstoff ist, ließ sich also eine Totalanalyse ausführen. Die gefundenen Werte ergeben addiert 100.19%, so daß ein Zweifel an der Zusammensetzung der Verbindung ausgeschlossen ist. Der verhältnismäßig niedrige Siedepunkt und die übrigen Eigenschaften der Säure zeigen, daß ihr wohl die einfache Formel $C_2H_4S_2$, nicht etwa eine polymere, zukommt.

Das spezifische Gewicht der Verbindung beträgt bei 20° 1.24.

Sie bildet mit den Alkalien völlig neutrale wasserlösliche Salze und ließ sich daher scharf titrieren.

0.1074 g Sbst. wurden in einem Kugelröhrchen abgewogen in eine abgemessene Menge 1/10-n. Kalilösung gebracht und unter der Flüssigkeit das Röhrchen zertrümmert. (Trotz dieser Vorsichtsmaßregel machte sich der entsetzliche Geruch der Säure momentan bis in weitere Entfernungen geltend.) Dann wurde mit Wasser verdünnt, bis die Farbe sich stark aufgehellt hatte, ein Tropfen Phenolphthaleinlösung zugesetzt und mit 1/10-n. Schwefelsäure bis zum Verschwinden der Rötung neutralisiert. Der Umschlag ist völlig scharf und führt zu einer weingelben Lösung. Es wurden so verbraucht 11.61 ccm 1/10-n. Kalilauge. Auf eine einbasische Säure der Formel $C_2H_4S_2$ berechnen sich 11.65 ccm.

Die geschwefelte Essigsäure ist eine sehr starke, in der Affinität zu Basen ihr sauerstoffhaltiges Analogon übertreffende Säure. Dies kann nicht wundernehmen, wenn man bedenkt, daß, auch die Mercaptane den entsprechenden Alkoholen gegenüber stärker saure Natur zeigen. Ebenso ist ja auch der Schwefelwasserstoff in dieser Richtung dem Wasser überlegen, weshalb seinerzeit von K e k u l é in seiner Habilitationsschrift die mittlerweile allgemein angenommene These aufgestellt werden konnte: »Der Schwefelwasserstoff ist eine Säure.« Wahrscheinlich übertrifft schon die K e k u l é - sche Thiacetsäure die Essigsäure in der Affinität merkbar. Ebenso wirkt anscheinend der CH_3.CS-Rest stärker acidifizierend als die Acetylgruppe. Die Methylcarbithiosäure vermag daher die Essigsäure aus ihren Salzen zu verdrängen. Gibt man sie zu einer Natriumacetatlösung, so färbt sich die Flüssigkeit sehr bald gelb unter Bildung von methylcarbithiosaurem Natrium und freier Essigsäure. Auf Zusatz von Salzsäure fällt die Säure in feinen Tröpfchen wieder aus. Auch ameisensaure Salze wie Kaliumformiat scheinen durch die Methylcarbithiosäure zerlegt zu werden, obschon die Ameisensäure eine etwa 12-mal stärkere Säure als Essigsäure ist. Jedenfalls besitzt auch verdünnte Alkaliformiatlösung ein weit größeres Lösungsvermögen für die Methylcarbithiosäure als Wasser, färbt sich damit sofort gelb und scheidet auf Zusatz von Mineralsäure die freie Carbithiosäure wieder aus. Destilliertes Wasser nimmt dagegen erst beim Erwärmen unter schwacher Gelbfärbung etwas der Säure auf und scheidet sie beim Erkalten teilweise als Trübung aus.

Mit Wasserdämpfen ist die Methylcarbithiosäure leicht flüchtig.

Um die Säure mit den Carbonsäuren, speziell der Essig- und Ameisensäure, bezüglich ihrer Stärke etwa vergleichen zu können, wurde versucht, ihre Dissoziationskonstante zu bestimmen. Sie oxydiert sich indessen so leicht, daß wir davon Abstand nehmen mußten. Beim Schütteln mit Wasser zur Herstellung einer verdünnten Lösung bildeten sich bald Klümpchen einer festen, gelben Verbindung, die vermutlich aus Thioacetyldisulfid bestand.

Obschon die Methylcarbithiosäure ca. 70% Schwefel enthält, also beinahe flüssigen Schwefel vorstellt, ist sie überraschenderweise nicht leicht zu entflammen. Sie kann mit einer Bunsenflamme berührt werden, ohne sich zu entzünden. Erst beim Erwärmen und Verdampfen verbrennt sie mit blauer Flamme unter Entwicklung von Schwefeldioxyd und Hinterlassung schwer verbrennlicher Kohle.

Papier und andere organische Stoffe werden von der Substanz intensiv rotgelb gefärbt. Auf der Haut erzeugt sie unverwüstliche schwarze Flecken.

Abkühlung mit Eis-Kochsalz brachte sie ebensowenig zum Erstarren wie rasch verdampfender Äther.

Die Alkali-, Erdalkali-, Aluminium- und Magnesiumsalze der Methylcarbithiosäure sind wasserlöslich. Eine neutrale, durch Titration erhaltene Lösung von methylcarbithiosaurem Natrium gab mit Bleiacetat hellgelbe, mit Sublimatlösung gelbe Fällung. Silbernitrat und Kupfersulfat gaben rotbraunen, Kobaltnitrat braunschwarzen, Nickelsulfat dunkelgrünen, Cadmiumsulfat gelben Niederschlag. Eisenchlorid lieferte eine braune, Ferrosulfat eine braunrote Fällung, die beide leicht von Äther wie von Chloroform gelöst wurden.

Leitet man trocknen Chlorwasserstoff in die Säure, so verwandelt sie sich bald in eine zähe, gelbe Masse, die beim Schütteln mit Äther in gelbliche Flocken übergeht.

Mit Jod-Jodkalium-Lösung entsteht ebenso wie bei der Einwirkung der Luft das Thioacetyldisulfid als gelber Niederschlag.

In alkalischer Lösung reduziert, gibt die Säure ein gelbliches, geruchloses Öl.

COMPARISON OF THE GERMANIC LANGUAGES. As was done with the Latin group, the Sabatier sample text has been translated into each of these languages, to enable the chemist to compare them with each other and with English. These translations were done by chemists each writing in his mother tongue.

The results are much the same as with the Latin group. On account of the scarcity of chemical articles in Norwegian and of the close similarity of the language to Danish no article in it is examined.

The five Germanic languages are very similar, as will be seen by a study of the Germanic glossary. In many cases a German dictionary is of considerable help with the other four related languages. As in the Latin group, chemical and international words, and words cognate with English run through all. Figures on these are given below:

Recognizable Words	German	Dutch	Swedish	Norwegian	Danish
Chemical	5	5	5	7	5
International	6	7	6	7	6
Cognate	1	3	7	6	6

Comparing the Latin and Germanic groups, we see that the number of chemical words is nearly the same. The number of international words is appreciably lower in the Germanic than in the Latin group, though these words are still sufficiently numerous to help the chemist considerably.

The Catalytic Hydrogenation of Acetylene

Dutch — Translated by D. Twiss

Acetyleen verbindt zich met waterstof in de koude in tegenwoordigheid van platinum zwart, eerst gevende ethyleen dan ethaan.

In tegenwoordigheid van een overmaat van waterstof, wordt acetyleen geheel getransformeerd in zuivere ethaan zonder eenige byreacties.

By 180° dezelfde reactie verloopt vlugger, maar er is vorming van een zekere hoeveelheid hoogere vloeibare koolwaterstoffen. Door vergrooting van de verhouding van acetyleen in het mengsel wordt ethyleen het voornaamste product maar wat ethaan wordt altyd gevormd niettegenstaande wat onveranderd acetyleen overblyft.

Wanneer de verhouding van acetyleen groot genoeg wordt, met platinum zwart by 180°, een zekere hoeveelheid koolstof wordt uit het gas afgescheiden hetgeen leidt tot gloeien van het platinum zooals het geval is wanneer nikkel wordt gebruikt.

Platinum spons is niet actief by lage temperatuur en heeft geen effekt op de hydratie van acetyleen behalve boven 180°.

Swedish — Translated by S. Sunner

Acetylen reagerar med väte vid rumstemperatur i närvaro av platinasvart och ger först etylen sedan etan.

I närvaro av överskott av väte omvandlas acetylen fullständigt till ren etan helt utan sidoreaktioner.

Vid 180° sker samma reaktion snabbare, men en viss mängd av högre, flytande kolväten bildas samtidigt. Genom att öka andelen acetylen i blandningen blir etylen huvudprodukten, men en viss mängd etan bildas alltid även om oförändrad acetylen återstår.

Om acetylenmängden ytterligare ökas, observerar man en svag sönderdelning under rökutveckling, som avslutas med att katalysatorn blir glödande på samma sätt som när nickel användes.

Platinasvamp är inte aktiv vid rumstemperatur och hydrerar inte acetylen förrän vid temperaturer över 180°.

Danish — Translated by J. Carstensen

Acetylen forbinder sig med brint i kulden i tilstedeværelse af platinsort, i begyndelsen former det ætylen, senere ætan.

I tilstedeværelse af overskud af brint vil acetylen fuldstændigt omdannes til rent ætan uden bireaktioner.

Ved 180° går reaktionen hastigere, men her forekommer der en vis mængde flydende højere kulbrinter. Ved forhøjelse af acetylen procenten i blandingen vil ætylen blive hovedproduktet, men noget ætan formes altid selvom ureageret acetylen er tilstede.

Hvis acetylen indholdet bliver stort nok (med platinsort ved 180°), vil der forekomme en vis mængde rygende dekomponeret luftart, og det ender med inkandescens, ligesom det er tilfældet med nikkel som katalysator.

Platinsvamp er ikke virksomt ved stuetemperatur og er ikke virksomt overfor brintning af acetylen, undtagen over 180°.

German — Translated by B. Finkelstein

Acetylen verbindet sich in Berührung mit Platinmohr in der Kälte mit Wasserstoff erst zu Äthylen, dann zu Äthan.

In Gegenwart überschüssigen Wasserstoffs verwandelt sich das Acetylen quantitativ in reines Äthan.

Bei 180° verläuft dieselbe Reaktion mit grösserer Geschwindigkeit, es bilden sich aber daneben flüssige höhere Kohlenwasserstoffe. Steigert man die Konzentration des Acetylens in dem Gasgemisch, so wiegt das Äthylen unter den Reaktionsprodukten vor, aber es entsteht trotzdem immer Äthan, selbst wenn ein Teil des Acetylens unverändert bleibt.

Steigt die Acetylenkonzentration über eine gewisse Grenze, so tritt bei 180° eine Zersetzung unter Kohlenstoffabscheidung ein, die schliesslich unter Aufglühen des Platins zu ähnlichen Erscheinungen führt, wie sie Nickel hervorruft.

Platinschwamm ist in der Kälte wirkungslos; erst oberhalb von 180° vermag er die Hydrierung des Acetylens herbeizuführen.

Norwegian — Translated by J. Jensen

Acetylen forbinder seg med hydrogen ved romtemperatur i nærvær av platina-sort og gir først etylen, dernest etan.

I nærvær av overskudd av hydrogen omdannes acetylen fullstendig til ren etan uten noen sidereaksjoner.

Ved 180°C foregår samme reaksjon fortere, men der er samtidig dannelse av en viss mengde høyere, væskeformige hydrokarboner. Ved å øke den relative mengde acetylen i blandingen, blir etylen hovedproduktet, men noe etan blir også dannet selvom der foreligger uomsatt acetylen.

Dersom den relative mengde acetylen blir stor nok (platina-sort ved 180°C), observeres røk-formige spaltningsprodukter av gassen. Man får tilslutt glødning på samme måte som med nikkel som katalysator.

Platina-svamp er ikke aktiv ved romtemperatur, og forårsaker ikke hydrogenering av acetylen unntatt over 180°C.

English — Translated by E. E. Reid

Acetylene combines with hydrogen in the cold in the presence of platinum black, giving first ethylene and then ethane.

In presence of an excess of hydrogen, acetylene is entirely transformed into pure ethane without any side reactions.

At 180° the same reaction takes place more rapidly but there is the formation of a certain amount of higher liquid hydrocarbons. By augmenting the proportion of acetylene in the mixture, ethylene becomes the main product but some ethane is always formed even though unchanged acetylene remains.

If the proportion of the acetylene becomes great enough, with the platinum black at 180°, a certain amount of smoky decomposition of the gas is observed and this ends with incandescence, as is the case with nickel.

Platinum sponge is not active in the cold and does not effect the hydrogenation of acetylene except above 180°.

LANGUAGES UNRELATED TO ENGLISH

In dealing with chemical articles in these languages the chemist misses cognate words but can get on without them as was demonstrated in the examination of the Polish article. Three more languages are considered: Hungarian, Finnish, and Czech. The first two, though of remote common Finno-Ugric origin, are so different from each other and from other languages that they have to be treated separately. The third one of these is closely related to Polish and Russian.

HUNGARIAN

Hungarian is a unique language which is very difficult to master. Its nearest relative is Finnish, but the relationship is strictly grammatical and is not reflected by the vocabulary. If a Hungarian and a Finn met accidentally, they would be unable to communicate. The only thing that is easy in the Hungarian language is the phonetic spelling.

Hungarian is a nonprepositional language, i.e., the prepositions are replaced by endings which are attached to the nouns and their exact form will vary with the vowel content of the nouns. The Hungarian equivalent of "in the room" is *a szobában*, while the Hungarian translation of "in the hall" is *a teremben*. Similarly, *az asztalon* means "on the table" and *a képen* "on the picture."

There are personal pronouns but they are almost never used with verbs as the ending of a verb indicates the person, such as:

I study	tanul*ok*
you study	tanul*sz*
he studies	tanul
we study	tanul*unk*
you study	tanul*tok*
they study	tanul*nak*

Again, these endings vary with the verbs, tenses, modes, etc.

He, *she*, and *it* are all expressed by *ö*. One is supposed to know from the context whether the statement refers to a man, woman, or object. Thus it is to be understood that no offense is meant when a Hungarian-American alters your sex in a conversation.

The number of grammatical rules is exceeded only by the exceptions to these rules. Considering all these complexities it is hard to believe that the young children playing in the parks of Budapest are speaking a perfect Hungarian. Luckily, chemists will seldom be put to the test of reading a Hungarian scientific paper in the original since almost every important Hungarian scientific publication is translated into German and published in a pertinent German journal.

The experimental section of an article by László Szekeres (*Magy Kem Folyoirat* 59: 228–29 [1953]; *CA* 48: 11347) on the preparation of thiocarbamic esters is given below:

Kísérleti rész

ω-Rodán-acetofenon, (C_9H_7ONS)

2,2 g KSCN és 4 g fenacil-bromidból, 50 ml etilalkoholban forralással. Termelés 2,7 g, fehér kristályok. O.p. = 75–76°.

Tiokarbaminsav-S-fenacilészter, ($C_9H_9O_2NS$). 20 g rodán-acetofenont 100 ml absz. dioxánban szuszpendáltam, majd 12 órán keresztül lassú áramban sósavgázt vezettem az elegybe, gondoskodván arról, hogy az elegy nedves levegővel ne érintkezhessék. A dioxános elegy lassan kissé megszínesedett, s a rodán-acetofenon néhány óra alatt szobahőmérsékleten feloldódott. A sósavgáz-bevezetés megszüntetése után kalciumkloridos csővel elzárva 6 napig állni hagytam az elegyet, majd erős keverés közben 1000 ml vízbe öntöttem. A hígítás hatására gyorsan kristályokká dermedő olaj vált ki, amit néhány órai állás után tégelyen szűrtem, vízzel mostam és 40°-on szárítottam. 17 g kissé sárga, 53–55°-on olvadó kristályokat nyertem. A termék petroléterből átkristályosítva 55–57°-on olvadó, kissé sárgás árnyalatú fehér, vattás szerkezetű, könnyű kristályokat eredményez.

$C_9H_9O_2NS$-re számított: C = 55,38%, H = 4,61%, N = 7,18%, S = 16,4%,

talált: C = 54,94%, H = 4,67%, N = 7,02%, S = 16,04%.

2-Oxi-4-fenil-tiazol, (C_9H_7ONS)

1 g 55–57°-on olvadó anyagot 25 ml etanol és 25 ml 35%-os sósav elegyével visszafolyó hűtő alatt három órán keresztül forraltam. Másnapra az oldatból kristályok váltak ki. A kivált kristályokat tégelyen szűrtem, 60–70°-on szárítottam. 0,57 g 195°-on olvadó, barna, tűs kristályokat nyertem. A terméket 20 ml benzolból átkristályosítva 0,31 g 204–205°-on olvadó, fehér kristályokat nyertem. A termék *Arapides* módszerével készített 2-oxi-4-fenil-tiazollal olvadáspont-depressziót nemmutat.

Talált: C = 61,46%; H = 4.21%; N = 7,68%
Számított: C = 61,75%; H = 3,96%; N = 7,92%

Following the recognizable words and with some help from the dictionary we have:

ω-Rodán-acetofenon, (C_9H_7ONS)

2,2 g KSCN . . . 4 g fenacil-bromidból, 50 ml etilalkoholban . . . 2,7 g, . . . kristályok. O.p (m.p.) = 75–76°.

Tiokarbaminsav-S-fenacileszter, ($C_9H_9O_2NS$)

20 g rodán-acetofenont 100 ml absz. dioxánban szuszpendáltam, . . .

12 órán . . . sósavgázt [HCl gas] . . . . dioxános . . . rodánacetofenon . . . óra . . . . sósavgáz- . . . kalciumkloridos . . . 6 napig [days] . . . 1000 ml vízbe [water] öntöttem [poured into]. . . . kristályokká . . . órai . . . vízzel mostam [washed] . . . 40°-on szárítottam [dried]. 17 g . . . 53–55°-on olvadó [melt] kristályokat . . . . petroléterből átkristályosítva 55–57°-on olvadó, . . . kristályokat . . . .

The rest is left as an exercise.

FINNISH — *by J. Bjorksten*

Finnish is related to Hungarian about as closely — or distantly — as German is to Spanish: Their words have little in common but the grammars have similarities such as the use of multiple infinitives and the use of suffixes or cases instead of prepositions which do not exist in Finnish. In declension of nouns there are fifteen cases formed by changes at the end of the word and these express things that would need prepositions in English. Finnish has no grammatical genders; it uses several infinitive forms which, together with cases, often make it possible to express by a single inflected word what would require a clause in a Caucasian language. This is confusing to anyone trying to decipher Finnish with a dictionary, because words can be so highly modified by these various cases, suffixes, and infinitives as to become unrecognizable. For example:

sakka	precipitate
saostaa	to precipitate
sakkana	in the capacity of a precipitate
saostettuaan	after he had precipitated
saostakseen	in order for him to precipitate
saostettuani	after I had precipitated
sakalta	from the top of the precipitate
sakasta	from the inside of the precipitate
saostama	precipitated by
sakaksi	transformation into a precipitate
sakan	the entire precipitate
sakkaa	part of a precipitate
sakkaan	into the precipitate

Or to take a verb:

tehdä	to make
teen	I make
tehtyään	after he had made
tekemä	made by
tekemällä	by making
tekemätön	not made
tekemään	in order to make
tekee	a third person makes (Finnish cannot distinguish between he and she)

The Finnish language is characterized by a very high ratio of vowels to consonants and by the non-existence in this language of the consonants b, c, f, g, w, x, and z.

Those chemists who enjoy a challenge and are not deterred by intricacies of Finnish are given here a chance to try their skill on a part of the experimental portion of a Finnish article on the bromination of toluene and cumene by Väinö Veijola (*Suomen Kemistilehti* 26A: 272–74 [1953]; *CA* 51: 8666):

1. Tolueeniin lisätty bromi reagoi, melkein riippumatta tolueenin bromi-konsentraatiosta, huomattavasti nopeammin kuin tolueeniin liuennut bromi. Kyseisissä olosuhteissa bromista reagoi lisättäessä noin 15% lisätystä bromista laskien ja 2,1–1,5% tolueeniin liuenneesta bromista laskien. Tolueeniin liuenneesta bromista sensijaan reagoi ainoastaan noin 0,25% bromin väkevyyden ollessa 1 mmol. Br_2/cm^3 tolueenia.

2. Analyysien mukaan bromautuminen tapahtui pääasiassa runkosubstituutiona. Sivuketjuun meni vain noin 3% bromista 30:ssa°C.

Bromaus valossa. Bromaus suoritettiin sekä päivänvalossa, jolloin huoneen valoisuus mitattiin luxeina, että käyttäen Hg-lamppua, jossa ultravioletti alue oli hyvin pieni.

1. Huoneen valoisuuden ollessa alle 100 luxia ei valon voimakkuus riittänyt täysin aktivoimaan tolueenin bromausta. Kun valoisuus kasvoi 120:een Luxiin, reagoi kaikki bromi, vähäistä induktioperiodia alussa lukuunottamatta sitä mukaa kuin sitä reaktioastiaan syötettiin. Tällöin tolueenin väri pysyi kuitenkin bromin värisenä koko reaktion ajan. Sivuketjuun meni bromista noin 87%. Hg-lamppua käytettäessä sivuketjuun meni noin 80–90%, riippuen jossakin määrin lampun etäisyydestä ja tolueeni-bromi-suhteesta.

2. Kumeenin bromausta aktivoimaan riitti huoneen valoisuus 5 luxia. Huoneen valoisuuden kasvaessa 5:stä 120:een luxiin pysyi sivuketjubromin määrä melkein vakiona, 84–85%:na. Hg-lampun käyttö ei lisännyt sivuketjun osuutta kokonaisbromista.

3. Valon voimakkuudella oli aivan ratkaiseva vaikutus bromautumisen kohdistumisessa isopropyyliryhmän eri hiiliin. Huoneen valoisuuden kasvaessa 5:stä 120:een luxiin α-bromin määrä väheni 62%:sta 33%:iin. Hg-lampun valo edisti β-hiilien bromautumista niin voimakkaasti, että α-bromin osuus sivuketjubromista oli vain noin 16–18%.

In the following examination the meanings of words taken from the Finnish glossary in the appendix are designated by *g* and those taken from the dictionary by *d*:

1. Tolueeniin lisätty [added to — *g*] bromi reagoi, melkein [nearly — *g*] riippumatta [regardless of — *d*] tolueenin bromikonsentraatiosta, huomattavasti (considerably — *g*) nopeammin [quicker — *g*] ... tolueeniin liuennut [dissolved — *g*] bromi. ... olosuhteissa [conditions — *d*] bromista reagoi lisättäessä noin [about — *g*] 15% lisätystä bromista ... 2,1–1,5% tolueeniin liuenneesta bromista ... Tolueeniin liuenneesta bromista sensijaan [on the other hand — *d*] reagoi ... noin 0,25% bromin ... 1 mmol. Br_2/cm^3 tolueenia.

2. Analyysien ... bromautuminen tapahtui [occurs — *d*] pääasiassa [mainly — *d*] runkosubstituutiona [ring substitution — *d*]. Sivuketjuun [side chain — *d*] meni vain [less than — *g*] noin 3% bromista 30:ssa °C.

Bromaus valossa. [in the light — *d*] Bromaus suoritettiin [perform — *d*] sekä [among — *d*] päivänvalossa [sunlight — *d*], jolloin [when — *d*] huoneen [room — *d*] valoisuus mitattiin [measured — *g*] luxeina, ... käyttäen [used — *g*] Hg-lamppua, jossa [where — *d*] ultravioletti alue [region — *d*] ... hyvin [very — *d*] pieni [low — *g*].

The rest of the article is left as an exercise.

The Catalytic Hydrogenation of Acetylene

Hungarian — Translated by I. E. Berck

Az acetilén hidegen vegyül hidrogénnel platina korom jelenlétében, előbb etilént, majd etánt adva.

Fölös hidrogén jelenlétében az acetilén minden melléktermék nélkül teljesen tiszta etánná alakul át.

180° hőmérsékleten ugyanez a reakció gyorsabban megy végbe, de bizonyos mennyiségű magasabb rendű folyékony szénhidrogének is képződnek. Ha a keverékben az acetilén arányát növeljük, etilén lesz a főtermék, de mindig képződik valamennyi etán, még akkor is ha az acetilén egy resze változatlan marad.

Ha a változatlan acetilén mennyisége elég nagy, minthogy a platina korom 180° hőmérsékleten van, bizonyos mérvű kormos gáz elbomlás észlelhető, amely végül a nikkelhez hasonlóan izzást eredményez.

Platina szivacs hidegen hatástalan és csak 180° hőmérséklet fölött katalizálja az acetilén hidrogénezését.

Finnish — Translated by E. Tommila

Asetyleeni yhtyy vetyyn kylmässä platinamustan läsnäollessa muodostaen ensin etyleeniä ja sitten etaania.

Vety-ylimäärässä asetyleeni muuttuu täydellisesti etaaniksi ilman sivureaktioita.

180°:ssa tämä reaktio tapahtuu nopeammin, mutta samalla muodostuu korkeampia nestemäisiä hiilivetyjä. Lisättäessä asetyleenin osuutta kaasuseoksessa etyleenistä tulee päätuote, mutta jonkin verran etaania syntyy silti aina, vaikka osa asetyleeniä jäisi muuttumattakin.

Jos asetyleenin osuus nousee kyllin suureksi, 180°:ssa tapahtuu platinan alkaessa hehkua hajaantumista, joka tuottaa nokea kuten nikkeliä käytettäessä.

Platinasieni on kylmässä tehoton; vasta 180°:n yläpuolella se voi aikaansaada asetyleenin hydrautumista.

English — Translated by E. E. Reid

Acetylene combines with hydrogen in the cold in the presence of platinum black, giving first ethylene and then ethane.

In presence of an excess of hydrogen, acetylene is entirely transformed into pure ethane without any side reactions.

At 180° the same reaction takes place more rapidly but there is the formation of a certain amount of higher liquid hydrocarbons. By augmenting the proportion of acetylene in the mixture, ethylene becomes the main product but some ethane is always formed even though unchanged acetylene remains.

If the proportion of the acetylene becomes great enough, with the platinum black at 180°, a certain amount of smoky decomposition of the gas is observed and this ends with incandescence, as is the case with nickel.

Platinum sponge is not active in the cold and does not effect the hydrogenation of acetylene except above 180°.

COMPARISON OF THE FINNO-UGRIC LANGUAGES. To show the similarities and dissimilarities of these languages, versions of the same Sabatier sections are given in Hungarian and Finnish. The Polish and Czech versions are in Chapter 3 to stress their similarity to Russian.

Although these languages are far removed from English it is found that a chemist can contrive to get information out of an article provided he is familiar with the subject.

As will be noted in Chapter 3, the Polish and Czech versions of the Sabatier section are very similar. It is, however, quite otherwise with the Hungarian and Finnish. The only thing they have in common is the presence of a number of the same chemical and international words:

Recognizable Words	Hungarian	Finnish
Chemical	6	5
International	2	1
Cognate	2	1

CZECH

The Czech language is closely related to Polish and to Russian. Nearly half of the Czech words in the glossary are recognizable from their resemblance to Polish. Czech has preserved the characteristics of an inflectional tongue. There are seven cases and three grammatical genders. The word order is practically the same as in English. The pronunciation follows the phonetical spelling; some of the letters with diacritical marks have sounds peculiar to the language.

The experimental section of an article on oxidation of cumene derivatives by J. Šlosar and V. Štěrba (*Chem. Průmysl* 15(4): 206–9 [1965]; *CA* 63: 4195) is chosen for examination:

Pokusná část

Příprava halogenderivátů kumenu

Chlórkumen: Do 5litrové tříhrdlé baňky byly nasazeny 2 l chlórbenzenu a 400 ml konc. H_2SO_4 a při 35 °C byl uváděn za míchání z tlakové láhve propylén (celkem 115 l). Za 2 hodiny byla kyselina sírová oddělena, organická vrstva ještě dvakrát promyta konc. kyselinou sírovou a potom nasyceným roztokem chloridu sodného. Dále byla organická vrstva 1 hodinu míchána s tuhou potaší při 50–60°C, zfiltrována a destilována na koloně ve vakuu. Nejprve byl vydestilován nezreagovaný chlórbenzen, potom chlórkumenová frakce v rozmezí 78 až 82°C při 15 torrech. Výtěžek byl 402 g, tj. asi 55% na propylén, uvedený do reakční směsi. Získaný produkt obsahoval asi 60–70% *o-* a 30–40% *p*-chlórkumenu. 200 g destilátu bylo rektifikováno za vakua 20 torrů na koloně o 30 teoretických patrech s refluxním poměrem 1:10. Byly získány 2 frakce, prvá (asi 40%) obsahovala převážně *o*-chlórkumen, druhá frakce obsahovala 90–95% *p*-chlórkumenu. Poloha izopropylskupiny byla zjištěna jednak plynovou chromatografií srovnáním s autentickými vzorky *o-* a *p*-chlórkumenu, jednak oxidací kyselinou dusičnou. Prvá frakce se prakticky neoxidovala, druhá poskytla v 70%ním výtěžku kyselinu *p*-nitrobenzoovou.

Taking the recognizable words we have the following:

Příprava halogenderivátů kumenu

Chlórkumen: ... 5litrové tříhrdlé baňky [three-necked flask, from context] ... 2 l chlórbenzenu ... 400 ml konc. H_2SO_4 ... 35°C ... propylén ... 115 l [propylene bubbled in]. ... 2 hodiny [hours, from dictionary] ... organická ... [organic layer separated, from context] ... promyta [washed, from context] konc. kyselinou [acid, from dictionary] sírovou [sulfuric, from dictionary] ... nasyceným [saturated, from dictionary] roztokem [solution, from context] chloridu sodného. ... organická vrstva 1 hodinu míchána [mixed mechanically, from context) ... tuhou potaší [this must be potassium carbonate, which would work best here] ... 50–60°C, zfiltrována ... destilována ... vakuu. ... vydestilován ... chlórbenzen, ... chlórkumenová frakce ... 78 ... 82°C ... 15 torrech (b_{15} 78–82°C). Výtěžek [yield, from dictionary] ... 402 g, ... 55% ... propylén, [55% based on propylene] ... produkt ... 60–70% *o-* ... 30–40% *p*-chlórkumenu.

The rest of the article is left as an exercise.

3

Russian

COMPARISON OF SLAVIC LANGUAGES

Chemists need not be appalled at the difficulty of gaining access to the huge amount of information being published in Russian chemical journals. The first step toward overcoming this particular language barrier has been taken in Chapter 2, where a Polish text was used to demonstrate the direct approach as a method of obtaining information from a foreign article. There is a close similarity between Russian, Polish, and Czech, although the latter two languages are printed in the Latin alphabet and Russian has one of its own. The resemblance of transliterated Russian to Polish and Czech is brought out by a comparison of the translations into the three languages of the Sabatier text.

The Russian is presented in transliterated form in order to nullify the effect of the difference in the alphabet. Acetylene, ethylene, and ethane run through all three versions, as do several international words. The close similarity of the languages is shown in the following tabulation:

English	Polish	Czech	Russian
hydrogen	wodorem	vodíkem	vodorodom
pure	czysty	čistý	chystyi
without	bez	bez	bez

An examination of the glossaries (see appendices) in which the three languages appear will show that in most cases the Polish, Czech, and transliterated Russian words are much alike. A Polish dictionary serves fairly well for Czech or transliterated Russian.

The Catalytic Hydrogenation of Acetylene

Polish — Translated by S. M. Newtowne

Acetylen łączy się z wodorem w obecności czarnej platyny w zimnie tworząc na początku etylen a potem etan.

W obecności nadmiaru wodoru acetylen jest całkowicie przetworzony w czysty etan bez żadnych ubocznych reakcji.

W temperaturze 180° ta sama reakcja odbywa się znacznie szybciej, ale również wytwarza się wtedy pewna ilość wyższych płynnych węglowodanów. Przy zwiększonej proporcji acetylenu w mieszance, etylen staje się głównym produktem, ale pewna ilość etanu zawsze wytwarza się, chociaż acetylen pozostaje niezmieniony.

Jeżeli proporcja acetylenu jest dostatecznie duża, w obecności czarnej platyny o temperaturze 180°, można zaobserwować pewną ilość dymu z dekompozycji gazu, co kończy się żarzeniem jak w przypadku z niklem.

Platynowa gąbka nie jest aktywna i nie wpływa na redukcję acetylenu w zimnie, a zaczyna działać tylko powyżej temperatury 180°.

Czech — Translated by M. Hudlický

Acetylen se slučuje s vodíkem za chladu za přítomnosti platinové černi nejprve na ethylen a pak na ethan.

Při přebytku vodíku se acetylen zcela přemění na čistý ethan bez vedlejších reakcí.

Při 180° nastává stejná reakce rychleji, ale vzniká jisté množství vyšších kapalných uhlovodíků. Zvýšením množství acetylenu ve směsi vzniká jako hlavní produkt ethylen, ale vždy se tvoří nějaký ethan, i když acetylen zcela nezreaguje.

Při dostatečně velkém přebytku acetylenu dochází za přítomnosti platinové černi při 180° k částečnému rozkladu za vzniku dýmu a posléze k rozžhavení podobně jako za použití niklu.

Platinová houba je za chladu neúčinná a při teplotě nižší než 180° nezpůsobuje hydrogenaci acetylenu.

Transliterated Russian — Translated by S. Trofimenko

Atsetilen soyedinyaetsa v kholode s vodorodom v nalichiyi platinovoi cherni i prevrashchayetsa snachala v etilen a potom v etan.

V nalichiyi izbytka vodoroda atsetilen polnostyu prevrashchayetsa v chystyi etan bez kakikhnebud pobochnykh produktov.

Pri 180°, reaktsiya proiskhodit bystreye, no poluchaetsa nekotoroye kolichestvo vyzshykh zhydkikh uglevodorodov. Kogda povyshat kolichestvennoye otnosheniye atsetilena v smesi, etilen dominiruyet mezhdu produktami, no etan vsegda prisutstvuet, dazhe yesli astayotsa nyeprevrashchonnyi atsetilen.

Yesli kolichestvo paslednyevo slishkom bolshoye, a platinovaya chern nagreta k 180°, zametno nekoye dymnoye razlazheniye gaza katoroye kanchayetsa vspyshkoi analogicheskoi k sluchayu nikela.

Platinovaya gubka astayotsa sovsem bez vliyaniya v kholodye i proyavlyayet aktivnost v hidrogenizatsiyi atsetilena tolko svyshe 180°.

INFORMATION FROM FORMULAE
AND NUMERALS ALONE

Actually, a chemist can get some information out of a Russian article even without knowing the alphabet, but how much depends on the chemist and the article.

Attention has been called above to the fact that a chemist, even without knowledge of the language, can get information out of an article from formulae, numerals, names of compounds, and international words. Just from formulae and numerals, which are printed in Russian exactly as they are in English, the chemist can learn something about what goes on. How much he will learn will depend somewhat on the article, but mainly on his knowledge of the subject and the effort which he makes. This is illustrated by the examination of a Russian article on the synthesis of homologs of cyclopentadiene by T. I. Naryshkina and I. F. Bel'skii (*Izv. Akad. Nauk SSSR Ser. Khim.* 3: 570–71 [1965]).

Экспериментальная Часть

2-Метил-5-этилфуран (I) получен в результате восстановления 2-метил-5-ацетилфурана в паровой фазе на скелетном Cu-Al-катализаторе [6]; т. кип. 117–118° (750 *мм*); n_D^{20} 1,4455, d_4^{20} 0,8930.

Гидролиз 2-метил-5-этилфурана проводился путем кипячения его с леляной уксусной кислотой и водой в течение 3 час. [3]. Из 110 *г* 2-метил-5-этилфурана, к которому было добавлено 165 *г* уксусной кислоты, 55 *г* дистиллированной воды и 2 капли серной кислоты, получили 86,1 *г* (63% от теорет.) гептандиона-2,5, т. кип. 84–87° (18 *мм*); n_D^{20} 1,4300; d_4^{20} 0,9563.

В реакцию циклизации брали 80 *г* гептандиона-2,5, 8 *г* едкого натра, 40 *мл* воды и 80 *г* метилового спирта. Смесь нагревалась в колбе с обратным холодильником в течение 10 час., затем продукт реакции экстрагировали эфиром и сушили сульфатом натрия. Остаток после отгона эфира перегоняли в вакууме. Выход 1,2-диметилциклопентен-1-она составлял 61,6 *г* (91% от теорет.); т. кип. 90–94° (30 *мм*); n_D^{20} 1,4900; d_4^{20} 0,9685.

1,2-Диметил-3-этилциклопентен-1-ол-3 был получен по реакции Гриньяра. К охлажденному до −15° эфирному раствору этилмагнийбромида (75 *г* C_2H_5Br и 19 *г* Mg) добавляли 56 *г* 1,2-диметилциклопентен-1-она-3. Получено 17,2 *г* (24% от теорет.) 1,2-диметил-3-этилциклопентен-1-ола-3 с т. кип. 60–62° (15 *мм*); n_D^{20} 1,4845; d_4^{20} 0,8481.

В результате дегидратации 15 *г* 1,2-диметил-3-этилциклопентанола-3 над сернокислым магнием при 310° и скорости пропускания 0,2 час⁻¹ собрано 12,1 *г* (92% от теорет.) углеводорода с т. кип. 40–42° (5 *мм*); n_D^{20} 1,4900; d_4^{20} 0,8497. Найдено: C 88,51; 88,70; H 11,68; 11,66%. Вычислено: C 88,53; H 11,47%.

Спектры снимались Е. В. Соболевым, за что авторы приносят ему искреннюю благодарность.

This example abounds in Geneva System names and complete structural formulae printed almost exactly as they would be in the *Journal of the American Chemical Society*. The formulae are grouped together in another part of the article and referred to in the experimental part by numerals. The following are the reactions that the chemist can learn from the structural formulae:

$$\begin{matrix} CH{:}CMe \\ | \\ CH{:}CEt \end{matrix} {>}O + H_2O \longrightarrow \begin{matrix} CH_2COMe \\ | \\ CH_2COEt \end{matrix}$$

The hydrolysis probably gives the dihydroxy heptadiene:

$$\begin{matrix} CH{:}C(OH)Me \\ | \\ CH{:}C(OH)Et \end{matrix}$$

which would isomerize to the heptadione. The next reaction is the self-condensation of 2,5-heptadione to the cyclic dimethyl pentenone:

$$\begin{matrix} CH_2COMe \\ | \\ CH_2COCH_2Me \end{matrix} \longrightarrow \begin{matrix} MeCCO \\ | \\ MeCCH_2 \end{matrix} {>}CH_2 + H_2O$$

The reaction of EtMgBr with a ketone and the dehydration of a secondary alcohol are well known:

$$\begin{matrix} MeCCO \\ \| \\ MeCCH_2 \end{matrix} CH_2 + EtMgBr \longrightarrow \begin{matrix} MeCC(OH)Et \\ \| \\ MeCCH_2 \end{matrix} {>}CH_2$$

$$\begin{matrix} MeCC(OH)Et \\ \| \\ MeCCH_2 \end{matrix} {>}CH_2 \longrightarrow \begin{matrix} MeCCEt \\ \| \\ MeCCH_2 \end{matrix} {>}CH + H_2O$$

From the structural formulae alone the chemist is able to write the reactions for all four steps. But, without recognizable international terms and not being able to look up words in the dictionary, he cannot learn just how these reactions were carried out.

Without knowing the Russian alphabet, the chemist can go only so far, but even so it is worthwhile to examine a Russian article. To eliminate the trouble caused by the strange alphabet, the Naryshkina/Bel'skii section is presented in transliterated form.

Eksperimental'naya Chast

2-Metil-5-etilfuran (I) poluchen v rezul'tate vosstanovleniya 2-metil-5-atsetilfurana v parovoi faze na skeletnom Cu-Al-katalizatore [6]; t. kip. 117–118° (750 mm); n_D^{20} 1.4455, d_4^{20} 0.8930.

Gidroliz 2-metil-5-etilfurana provodilsya putem kipyacheniya ego s ledyanoi uksusnoi kislotoi i vodoi v techenie 3 chas. [3]. Iz 110 g 2-metil-5-etilfurana, k kotoromu bylo dobavleno 165 g uksusnoi kisloty, 55 g distillirovannoi vody i 2 kapli sernoi kisloty, poluchili 86.1 g (63% ot teoret.) geptandiona-2,5, t. kip. 84–87° (18 mm); n_D^{20} 1.4300; d_4^{20} 0.9563.

V reaktsiyu tsiklizatsii brali 80 g geptandiona-2,5, 8 g edkogo natra, 40 ml vody i 80 g metilovogo spirta. Smes nagrevalas v kolbe s obratnym kholodil'nikom v techenie 10 chas., zatem produkt reaktsii ekstragirovali efirom i sushili sul'fatom natriya. Ostatok posle otgona efira peregonyali v vakuume. Vykhod 1,2-dimetiltsiklopenten-1-ona sostavlyal 61.6 g (91% ot teoret); t. kip. 90–94° (30 mm); n_D^{20} 1.4900; d_4^{20} 0.9685.

1,2-Dimetil-3-etiltsiklopenten-1-ol-3 byl poluchen po reaktsii Grin'yara. K okhlazhdennomu do −15° efirnomu rastvoru etilmagniibromida (75 g C_2H_5Br i 19 g Mg) dobavlyali 56 g 1,2-dimetiltsiklopenten-1-ona-3. Polucheno 17.2 g (24% ot teoret.) 1,2-dimetil-3-etiltsiklopenten-1-ola-3 s t. kip. 60–62° (15 mm); n_D^{20} 1.4845; d_4^{20} 0.8481.

V rezul'tate degidratatsii 15 g 1,2-dimetil-3-etiltsiklopentanola-3 nad sernokislym magniem pri 310° i skorosti propuskaniya 0.2 chas⁻¹ sobrano 12.1 g (92% ot teoret.) uglevodoroda s t. kip. 40–42° (5 mm); n_D^{20} 1.4900; d_4^{20} 0.8497. Naideno: C 88.51; 88.70; H 11.68; 11.66%. Vychisleno: C 88.53; H 11.47%.

Spektry snimalis E. V. Sobolevym, za chto avtory prinosyat emu iskren-nyuyu blagodarnost.

From recognizable words in the transliterated form we can get details of the operation:

2-Metil-5-etilfuran . . . rezul'tate [is — obtained from] . . . 2-metil-5-atsetil-furana . . . faze . . . skeletnom Cu-Alkatalizatore [the mention of copper-alumina catalyst suggests that this was by catalytic hydrogenation which is regularly effected in the vapor phase] . . .

Gidroliz 2-metil-5-etilfurana . . . uksusnoi kislotoi [acetic acid — from the transliterated Russian version of the Polish section in Chapter I] . . . 3 chas. [3 hours — *czas* is the Polish word for time] . . . 110 g 2-metil-5-etilfurana, . . . 165 g uksusnoi kisloty, 55 g distillirovannoi vody [*woda* in Polish is water] . . . 2 kapli [*kapla* is the Polish word for drop] sernoi kisloty, [sulfuric acid — *siarka* in Polish is sulfur] . . . 86.1 g (63% ot teoret.) geptandiona-2,5, . . .

. . . reaktsiyu tsiklizatsii . . . 80 g geptandiona-2,5, 8 g edkogo natra, [caustic soda is regularly used in this condensation] 40 ml vody . . . 80 g metilovogo spirta. [methanol] . . . 10 chas., [from the context we may surmise that the mixture is heated under reflux for 10 hours] . . . produkt reaktsii ekstragirovali efirom [ether] . . . sushili [*suchy* in Polish means dry] sul'fatom natriya. . . . efira peregonyali [peregrinate or wander off] . . . vakuume. . . . 1,2-dimetiltsiklopenten-1-ona . . . 61.6 g (91% ot teoret) . . .

1,2-Dimetil-3-etiltsiklopenten-1-ol-3 . . . reaktsii Grin'yara. . . . −15° efirnomu . . . etilmagniibromida (75 g C_2H_5Br . . . 19 g Mg) . . . 56 g 1,2-dimetiltsiklopenten-1-ona-3. . . . 17.2 g (24% ot teoret.) 1,2-dimetil-3-etiltsiklopenten-1-ola-3 . . .

. . . rezul'tate degidratatsii 15 g 1,2-dimetil-3-etiltsiklopentanola-3 . . . sernokislym magniem [magnesium sulfate] . . . 310° . . . skorosti [*skoro* in Polish means rapidly] . . . 0.2 chas⁻¹ . . . 12.1 g (92% ot teoret.) . . .

As was shown above, the chemist could get information on the reactions by following the structural formulae in the Russian version, but was prevented by the strange alphabet from learning anything further. With the help of transliteration, he is able to learn more, but to get complete information, the chemist needs the Russian alphabet. Until he learns it, he can sometimes obtain information from just a few clues. How to do this is shown by the examination of a Russian article that follows, on the synthesis of t-butylperoxide and hydroperoxide by A. E. Batog *et. al.* (*Ukr. Khim. Zh.* 31(2): 207-8 [1965]).

1. Синтез гидроперекиси третичного бутила. В автоклав из нержавеющей стали, содержащий 70%-ную серную кислоту, при $-5 \div +5°$ подавали из баллона изобутилен. Молярное отношение изобутилена к серной кислоте до 2:1. Затем смесь энергично встряхивали при охлаждении 2–3 часа до полного поглощения изобутилена.

Полученную смесь трет-бутилсерных кислот при энергичном перемешивании и температуре $-5 \div 0°$ прибавляли к 50%-ной перекиси водорода, взятой из расчета 1,05 моля на 1 моль поглощенного изобутилена, и продолжали перемешивание еще 2 часа. После полного разделения органический слой отделяли от кислотного, отмывали от остатков кислоты минимальным количеством 10%-ного раствора углекислого натрия и водой и сушили сульфатом натрия. Полученный алкилат содержит 70% гидроперекиси и 25% перекиси третичного бутила (по данным йодометрического определения). Выход перекисных соединений на поглощенный изобутилен 80–85%.

Для выделения гидроперекиси третичного бутила смесь обрабатывали при $0 \div 10°$ 20%-ным раствором едкого натра. После отделения перекиси трет-бутила раствор натриевой соли гидроперекиси промывали эфиром и подкисляли 70%-ной серной кислотой; гидроперекись отделяли, промывали раствором бикарбоната натрия и водой и сушили сульфатом натрия. Полученный продукт вакуумировали при 22–25°/15 мм. Остаток представляет собой гидроперекись третичного бутила. Т. кип. 41–42°/10 мм, n_D^{20} 1,3985, d_4^{20} 0,8943; MR_D 24,31; выч. 24,39. По лит. данным [2], т. кип. 35–37°/17 мм, n_D^{20} 1,4015, d_4^{20} 0,896.

2. Синтез перекиси третичного бутила. К трет-бутилсерной кислоте при энергичном перемешивании и температуре $5 \div 15°$ быстро приливали 50%-ную перекись водорода (0,5 моля на 1 моль поглощенного изобутилена) и продолжали перемешивание еще 8 час. После полного разделения органический слой отмывали от кислоты 10%-ным раствором углекислого натрия и водой и сушили сульфатом натрия. Выход перекиси 85%. Т. кип. 110–111°, 13°/20 мм, n_D^{20} 1,3900, d_4^{20} 0,7976, MR_D 43,45; выч. 42,98. По лит. данным [2], т. кип. 13–15°/20 мм, n_D^{20} 1,3910, d_4^{20} 0,7970.

In a previous preparation of the same compound (A. P. Meshcheryakov *et al.*, *Izv. Akad. Nauk SSSR, Otd. Khim. Nauk* [1955]: 742-49; *CA* 50: 2296) 63% H_2SO_4 was saturated with iso-butylene:

$$H_2SO_4 + Me_2C{:}CH_2 \longrightarrow t\text{-}BuO.SO_2.OH$$

The product was treated at 5° during one hour with 27% H_2O_2. After standing overnight the mixture of t-butylhydroperoxide and t-butylperoxide was separated, 80% of theoretical.

Having read this, the chemist can see that the preparation given in the Russian article is essentially the same except that 70% sulfuric acid had been used and that the hydrogen peroxide was 50% instead of 27%. He could have identified, from context, Russian words for isobutylene, percent, sulfuric acid, and hydrogen peroxide. Without knowing the alphabet he would have been unable to spot international words or to look up the Russian in the dictionary, and so could not get details of the preparation. This illustrates that the chemist who is well informed on the subject needs only a few clues to show him what is going on. However, he can go only so far without the knowledge of the alphabet.

Also, a chemist adapts his methods of getting information to the subject in hand. He would employ quite different methods for determining the identity and purity of samples of oils of mustard, cassia, and olive. In all cases, however, he would be guided by his acquaintance with the material in hand. The task of getting information may be easy or difficult, but in any case, the chemist must use his knowledge of the subject and his powers of deduction to get the maximum benefit from whatever clues he may find.

In this case the chemist was aided greatly by the fact that the article was one of a series on t-butylperoxide, so that all he had to do was to put it alongside one in the *Journal of the American Chemical Society* (N. A. Milas, S. A. Harris, *J. Am. Chem. Soc.* 60: 2434 [1938]) and note the similarities and differences. The chemist needed only one chemical term and two pairs of percentages, 70–63 and 27–50, as clues to show him how the present process differed from the one previously used. The article is here examined by a chemist sufficiently familiar with the alphabet to be able to spot the Russian forms of the following international and chemical terms: autoclave, isobutylene, energetic, temperature, organic, minimal, sodium, and steel.

With the aid of these and with the aid of the dictionary the chemist now attempts to get the complete story by taking the article sentence by sentence:

В автоклав из нержавеющей стали, содержащий 70%-ную серную кислоту, при −5÷+5° подавали из баллона изобутилен.

In an autoclave of stainless steel place 70% sulfuric acid at −5 to +5° and pass in isobutylene from a container. [Isobutylene boils at −6.9° and therefore must be kept in liquid form in a closed container.]

Молярное отношение изобутилена к серной кислоте до 2:1.

The molar ratio of isobutylene to sulfuric acid is 2:1.

Затем смесь энергично встряхивали при охлаждении 2–3 часа до полного поглощения изобутилена.

Thereupon the mixture is energetically shaken while cooling for 2–3 hours until absorption of isobutylene is completed. [This must mean that the absorption of isobutylene has gone as far as it can go. The sulfuric acid can react with only one mole of the two that were put in.]

Полученную смесь трет-бутилсерных кислот при энергичном перемешивании и температуре −5÷0° прибавляли к 50%-ной перекиси водорода, взятой из расчета 1,05 моля на 1 моль поглощенного изобутилена,
и продолжали перемешивание еще 2 часа.

The obtained mixture of tertiary butylsulfuric acid during energetic stirring at a temperature of −5 to 0° is added to 50% hydrogen peroxide,
calculated on 1.05 moles to 1 mole of absorbed isobutylene, and continue
stirring an additional two hours.

После полного разделения органический слой отделяли от кислотного, отмывали от остатков кислоты минимальным количеством 10%-
ного раствора углекислого натрия и водой и сушили сульфатом натрия.
Полученный алкилат содержит 70% гидроперекиси и 25% перекиси
третичного бутила . . .

After complete separation the organic layer is taken off from the acid,
washed free of residual acid by a minimal amount of 10% sodium carbonate
solution and with water and dried over sodium sulfate. The so obtained
products are 70% hydroperoxide and 25% tertiary butyl hydroperoxide.

Выход перекисных соединений на поглощенный изобутилен 80–85%.

The yield of peroxide calculated on the absorbed isobutylene is 80–85%.

THE RUSSIAN ALPHABET

The Russian alphabet consists of thirty-two letters, two of which are merely
markings to show how certain letters are pronounced. Some of the letters are
Greek, some Latin, and some of native Russian origin. The greek part is what
remains of the imported Greek alphabet after centuries of rough usage.

The Russian alphabet is given below with notes on the sounds and uses of
the letters, after each of which is placed a chemical name and an international
word in which the letter occurs:

А а   A a   алкил (alkil) alkyl
        (as in far)   магний (magnii) magnesium

Б б   B b   бром (brom) bromine
        (as in bench)   бензол (benzol) benzene

В в   V v   вакуум (vakuum) vacuum
        (as in vent)   вода (voda) water

Г г   G g   грамм (gramm) gram
        (as in go)   газ (gaz) gas
This letter is one of the few that can have a second pronunciation. One example:
сегодня (sevodnya). Also, transliterations of English into Russian generally substitute г for h: homolog—гомолог.

Д д   D d   димер (dimer) dimer
        (as in deep)   диполь (dipol) dipole

Е е   E e   гель (gel) gel
(as in ch*e*st)
An unstressed E-e is pronounced as in ch*e*st. A stressed E-e is *ye* as in *ye*s.

Ё ё   E e   лёд (lyod) ice
(as in *yo* of *yo*kel)

Ж ж   Zh zh   журнал (zhurnal) journal
(as in *s* of trea*s*ure)   жир (zhir) fat

З з   Z z   зона (zona) zone
(as in *z*ipper)   закон (zakon) law

И и   I i   ион (ion) ion
(as in mach*i*ne)   индекс (indeks) index

Й й   I i   йодометрический (iodometricheskii) iodometric
(as in mach*i*ne)   каустический (kausticheskii) caustic
This letter is И-и only slightly modified.

К к   K k   кетон (keton) ketone
(as in *k*ind)   кислота (kislota) acid

Л л   L l   лимит (limit) limit
(as in *l*og)   луч (luch) ray

М м   M m   масса (massa) mass
(as in tre*m*ble)   медь (med) copper

Н н   N n   негатив (negativ) negative
(as in *n*ut)   нефть (neft) petroleum

О о   O o   овал (oval) oval
(as in *o*ver)   окись (okis) oxide

П п   P p   продукт (produkt) product
(as in *p*en)   пиролиз (piroliz) pyrolysis

Р р   R r   реагент (reagent) reagent
(as in *r*un)   работа (rabota) work

С с   S s   сифон (sifon) syphon
(as in *s*ight)   сера (sera) sulfur

Т т   T t   тест (test) test
(as in *t*in)   тип (tip) type

У у   U u   курс (kurs) course
(as in t*u*ne)   пульс (pul's) pulse

Ф ф   F f   фаза (faza) phase
(as in *f*un)   фосфор (fosfor) phosphorus

Х х   Kh kh   хлор (khlor) chlorine
(as in *Kh*an)   химический (khimicheskii) chemical
This is a very harsh and guttural sound only roughly approximated here.

Ц ц   Ts ts   цикл (tsikl) cycle
(as in le*ts*)   центр (tsentr) center

Ч ч   Ch ch   часть (chast) part
(as in *ch*urch)

Ш ш   Sh sh   штат (shtat) state
(as in *sh*op)   шар (shar) sphere

Щ щ   Shch shch   щелок (shchelok) lye
      (as in wis*h* *ch*urch)

Ъ ъ   This is the "hard sign" and designates the pronunciation of other letters. It is omitted
      from the modern Russian alphabet.

Ы ы   Y y   мыло (mylo) soap
      (as in *i* of pr*i*sm)
      This is a very unusual sound and must be heard to be appreciated.

Ь ь   This is the "soft sign", also designating the pronunciation of other letters.

Э э   E e   эфир (efir) ether
      (as in *e*ver)   этан (etan) ethane

Ю ю   Yu yu   плюс (plyus) plus
      (as in *yu*le)   алюминий (alyuminii) aluminum

Я я   Ya ya   поляроид (polaroid) polaroid
      (as in *ya*cht)   яд (yad) poison

LEARNING THE ALPHABET

As the Russian alphabet bars access to half of the chemical articles in foreign languages, the chemist must learn it. The effort involved is but a small price to pay for the benefits it brings. Learning the alphabet is something which each individual has to do for himself, though he can be aided in various ways. Repetition and association are strongly recommended. Writing and repeating the letters many times helps fix them in the mind.

In a few evenings at home one chemist learned the letters well enough to spot international words and names of organic compounds and so was able to start going directly to Russian chemical articles. Of course, gaining facility requires practice.

In the examination of the Polish article on L-adrenaline in Chapter 2 the chemist was able to get the story mainly from ten chemical names and eight international words. Once he has learned the letters he can follow the same procedure in Russian articles. Being able to spot these names and words in a Russian chemical article, he is well on his way.

In an article in English we see "methyl" and "phenyl" as words without noting the individual letters. To scan a Russian article quickly the chemist must learn to recognize chemical names as complete words, in spite of their strange appearance. Actually the systematic names of organic compounds and the international words are practically the same as in English. They look strange in print, but are recognizable when they are pronounced.

To enable the chemist to gain speed in recognizing such chemical names and international words in unfamiliar garb, two lists are given in the Appendix, the first of chemical names and the second of international words. In both these lists the words are shown in the Russian alphabet, in transliterated form, and in English.

It is advisable to spend considerable time in going over these lists, in order to become able to spot the words quickly. This need not be done all at once. The words are remembered better if picked up one at a time, as the chemist scans one article after another. Work on these lists is an excellent way to acquire facility with the alphabet.

With the alphabet and with chemical names the chemist can work alone, but he can work more efficiently as a member of a small group. Assistance on pronunciation may be obtained from one acquainted with Russian or from a recording. Photostats of one or more Russian articles might be passed out to members of a group for individual examination. The results would be brought together for discussion or grading. According to the rules of the game, each chemist should familiarize himself with the subject of the article before undertaking to scan it. The glossary and dictionary would, of course, be used as needed. To aid the chemist, a glossary of three hundred and fifty frequently encountered Russian words is given in the Appendix.

SOME CHARACTERISTICS OF THE RUSSIAN LANGUAGE

The sentence structure in Russian is almost exactly the same as in English. On this account a chemist familiar with the alphabet has less difficulty with a Russian chemical article than with a German in spite of the absence of cognate words.

Russian is a highly inflected language. There are three genders and six cases making thirty-six endings if we count singular and plural. Let it suffice to say that nominative singular nouns (functioning as subjects) usually end in a consonant if they are masculine and the vowel я (a) if they are feminine.

Adjectival endings are much more easily classified. In general an adjective ends in ый, ий, or ой, if it is masculine; oe, ee, or óe if it is neuter; or aя, яя, or áя if it is feminine. Adverbs when formed from adjectives usually end in o.

Verbs also have an appalling variety of endings. For the present purpose, it is necessary to mention only basic endings for verbs of the first and third person. In the present tense, the usual ending for the first person singular verb is ю; first person plural is eм or им; and third person singular is eт or ит. In the past tense verbs agree in gender and number with their subject. For singular masculine verbs the ending is л; singular neuter is ло; singular feminine is ла; and all plural past tense verbs regardless of gender end in ли.

There is nothing in Russian corresponding to the English articles *a*, *an*, and *the*.

All of this makes learning to write Russian quite complicated, but none of it should bother a chemist in getting information out of a Russian chemical article. If he sees the words for *dilute sulfuric acid* and for *ferrous sulfide*, he knows what will happen if they are brought together and that it will make no difference whether a man or a woman does the mixing.

There are two peculiarities of Russian which might cause confusion. The verb "to be" is frequently omitted. Thus, one might see "petroleum — dark, oily liquid," instead of "petroleum is a dark, oily liquid." In English, as in many other languages, a double negative is a positive, but in Russian, as it was in Greek, a double negative is a negative.

COMPARISON OF RUSSIAN AND ENGLISH

Finally, to show the chemist what a familiar text looks like in Russian, a translation of the Sabatier section is given below:

The Catalytic Hydrogenation of Acetylene

Russian — Translated by S. Trofimenko

Ацетилен соединяется в холоде с водородом в наличии платиновой черни и превращается сначала в этилен, а потом в этан.

В наличии избытка водорода ацетилен полностю превращается в чистый этан без каких-нибудь побочных продуктов.

При 180° реакция происходит быстрее, но получается некоторое количество высших жидких углеводородов. Когда повышать количественное отношение ацетилена в смеси, этилен доминирует между продуктами, но этан всегда присутствует, даже если остается непревращенный ацетилен.

Если количество последнего слишком большое, а платиновая чернь нагрета к 180°, заметно некое дымное разложение газа которое кончается вспышкой аналогической к случаю никеля.

Платиновая губка остается совсем без влияния в холоде и проявляет активность в гидрогенизации ацетилена только свыше 180°.

English — Translated by E. E. Reid

Acetylene combines with hydrogen in the cold in the presence of platinum black, giving first ethylene and then ethane.

In presence of an excess of hydrogen, acetylene is entirely transformed into pure ethane without any side reactions.

At 180° the same reaction takes place more rapidly but there is the formation of a certain amount of higher liquid hydrocarbons. By augmenting the proportion of acetylene in the mixture, ethylene becomes the main product but some ethane is always formed even though unchanged acetylene remains.

If the proportion of the acetylene becomes great enough, with the platinum black at 180°, a certain amount of smoky decomposition of the gas is observed and this ends with incandescence, as is the case with nickel.

Platinum sponge is not active in the cold and does not effect the hydrogenation of acetylene except above 180°.

4

Japanese

The Japanese is similar to the Russian in that chemical names and international words are obscured by a strange alphabet. Able and industrious Japanese chemists are publishing thousands and thousands of chemical articles. They are considerate enough to print half of these in English and to append summaries to many others. This takes care of a part, but only a part of their publications. Some Japanese authors print systematic names of organic compounds in Latin type. We may hope that this will become general and that technical words will be treated in the same way. Inserting (H_2SO_4) and $(NaOH)$ and the like after the names of chemicals would be of much additional help.

Because of the immense number of Russian chemical articles, many chemists are learning Russian and huge sums are being spent by several agencies on massive translations. As this is not being done for Japanese articles, the chemist must find a way of handling them.

This chapter shows that the direct approach may be applied to Japanese chemical articles, as well as to those in other languages, and it contains some practical suggestions on how this may be done. On close examination, what appear to be insuperable difficulties turn out to be "paper tigers."

JAPANESE SYSTEMS OF WRITING

Japanese writers use three thousand Chinese ideographs and two alphabets, Katakana and Hirakana. Actually, these are two ways of writing the same letter, the Katakana corresponding to our printing and Hirakana to our script.

57

Hirakana is of little importance to chemists, as it is employed only occasionally for chemical terms. It is used principally for prepositions, verbal endings, and grammatical elements not essential to the scanner. There are Chinese ideographs for long-known ideas, materials, and processes, such as "atom," "iron," and "distillation." But "magnesium" and "vinyltriethoxysilane," with which Confucius was not familiar, are written in the Katakana alphabet. An ideograph represents a single idea so two or more of them may have to be combined to express a single word. Thus, "anhydrous" is written with two characters, one for "water" and the other for "without." The chemist may be confused with words of this type by the lack of spacing which would show where one word ends and another begins.

A large number of Japanese technical terms are transliterations of English and German into Katakana. Some words have been changed only slightly and others considerably as a consequence of linguistic differences. This is illustrated below:

English:	gas	magnesium	sodium	alcohol
Polish:	gaz	magnezja	sod	alkohol
Transliterated Russian:	gaz	magnii	natrii	alkogol
Transliterated Japanese:	gasu	maguneshûmu	natoryumu	arukôru

The basic alphabet has 48 symbols; in addition, there are modifications of sounds which double that number. As a guide to transliteration, the Katakana alphabet is given at the end of this chapter. The letters are arranged in five columns, each headed by a vowel. The English equivalent of each so-called letter is a syllable consisting of a consonant with the vowel at the head of the column. The sole exception is N.

SAMPLE TEXT ANALYZED

The approach to Japanese articles follows the usual pattern. From formulae and numerals which are printed exactly as in English, the chemist identifies the article and relates it to his own experience. He writes down what he believes to be the story. In checking and amplifying this, the Katakana words and ideographs must be handled in different ways. When transliterated, many of the Katakana words are found to be the names of compounds and international terms, usually immediately recognizable. The remainder are looked up in a dictionary of transliterated Japanese. The ideographs are taken care of by a method similar to that used in solving crossword puzzles. From the context the chemist thinks up words which might take the place of the ideographs and looks them up in an English-Japanese dictionary. He then compares the ideographs which he finds with those in the article. He may have to guess several times to find those that match. Just how these methods are used is illustrated by the examination of a Japanese article.

The one used here is on the synthesis of an organosilicon compound by Mareyoshi Momonoi and Niichiro Suzuki (Nippon Kagaku Zasshi 78: 1324–6 [1957]; *CA* 54: 5434g). The particular substance studied is vinylmethyldiethoxy-silane (III), which is prepared by the reaction of a methyl magnesium halide (II), a Grignard reagent, with vinyltriethoxysilane (I):

$$CH_2 = CH \cdot Si(OC_2H_5)_3 + CH_3MgX \longrightarrow \quad \begin{matrix} CH_2 = CH \\ \diagup \\ CH_3 \end{matrix} Si(OC_2H_5)_2 + Mg(OC_2H_5)X$$

$$\text{(I)} \qquad\qquad \text{(II)} \qquad\qquad\qquad \text{(III)}$$

Comparisons of the yields with different halides gives: I — 40%, Br — 65%, and Cl — 79%. Reference is made to a yield of 57.4%, X = Br, obtained by Cohen (M. Cohen, J. R. Ladd, *J. Am. Chem. Soc.* 75 [1953]).

The experimental section is reproduced on the following page:

1) ビニル-メチルジエトキシシラン (III)

イ) メチルマグネシウムクロリド (II; X=Cl)：四つ口フラスコにマグネシウム 36.4 g，脱水エーテル 300 g を仕込み，器内の空気を乾燥窒素ガスで置換する。これを 0°C に冷却してメチルクロリドを徐々に導入する。この間ヨウ素の小片を加え，臭化メチルを少量導入して反応を励起させる。マグネシウムがほとんど消失したら反応を中止してつぎの反応に用いる。

ロ) (II; X=Cl) と I との反応：四つ口フラスコに I 430 g を加え器内の空気を窒素ガスで置換したのち，かきまぜながらイ) で合成した (II; X=Cl) のエーテル溶液を分液漏斗から滴下反応させる。温度は 20°C 付近にたもつ。滴下終了後さらに 2 時間加熱還流してから放冷する。この反応液をロ過し，ロ液を蒸留する。bp 132°〜134°C の留分は III であり 177 g を得た。収率 79%，回収 I (bp 160°C) 165 g。

ハ) (II, X=Br) と I との反応：イ) と同じようにマグネシウム 24.3 g，脱水エーテル 400 g 中に臭化メチルを導入反応せしめて (II; X=Br) を合成し，I 250 g と反応させる。III の収量 98.5 g，収率 65.3%，回収 I 28.5 g。

ニ) (II; X=J) と I との反応：イ) と同じようにマグネシウム 24.3 g，脱水エーテル 400 g 中にヨウ化メチル 156 g を滴下反応させて(II; X=J) を合成し，I 268 g を反応させる。III の収量 77.9 g，収率 39.4%，回収 I 20 g。

RECOGNIZABLE FORMULAE. The first thing the chemist does is to look for recognizable formulae, numerals, or words. In the first two paragraphs he finds:

(III) ... (II; X = Cl) ... 36.4 g ... 300 g ... 0°C ... (II; X = Cl) ... I ... I 430 g ... (II; X = Cl) ... 20°C ... 2 ... bp 132° ∼ 134°C ... 177 g ... 79% ... I(bp 160°C) 165 g.

From these, in the light of his own experience with the Grignard reaction, the chemist writes down what he thinks was done. In a four-necked flask were placed 36.4 g of magnesium turnings and 300 g of anhydrous ether with a little iodine to initiate the reaction. Nitrogen may have been passed in to drive out the air. The flask was cooled in ice to 0°C and methyl chloride was bubbled in as a gas. The ether solution of methyl magnesium chloride is added slowly to a solution of I, vinyltriethoxysilane, keeping the temperature at 20°C. The resulting mixture was probably refluxed for an hour or two to insure the completion of the reaction. The ether was taken off and the residue fractionated giving 177 g III, vinylmethyldiethoxysilane, boiling at 132 ∼ 134°C, 79% yield, and 165 g of unreacted I, boiling at 160°C.

The chemist is pleased to learn of the efficiency of methyl magnesium chloride in this preparation. He is sure that he can duplicate the results even without knowing all the details of the procedure. However, it is desirable to check his conclusions and look for additional information. He starts by transliterating the Katakana according to the charts given at the end of this chapter: "biniru-mechiru-jietokishi-shiran ... mechirumaguneshûmukurorido ... Furasuko ... maguneshûmu ... êteru ... gasu ... mechirukurorido ... yôdo ... mechiru ... maguneshûmu ... furasuko ... gasu ... êteru ... ro ... ro."

Some of these words are immediately recognizable from their resemblance to English or German or from context. Others can be figured out with the aid of rules given in the introduction to the glossary. Thus, in êteru the t becomes th and the u is dropped giving "ether." Substituting l for r and leaving out the u's in furasuko gives flasko, which suggests "flask." The yôdo is recognizable as Jod, German for iodine. When iodine is looked up, it turns out to be partly ideograph. To obtain the meanings of the two words consisting of ro and an ideograph, the chemist looks at the words in the dictionary beginning with ro and finds "filter" and "filtrate."

IDEOGRAPHS

As far as they go, the transliterated Katakana words substantiate the chemist's story, but leave some blanks. The chemist looks up in an English-Japanese dictionary words which might fill these and checks the ideographs with those in the text. If the chemist's first guess is not correct, he must keep guessing until he finds the word which fits. For example, he had supposed that the container was the usual three-necked flask. The word qualifying flask should be "necked," but this was not found. After several guesses, he hit upon "hole," the character for which matched the character before "flask." The numeral preceding "flask" was not three as he had assumed, but four. As nitrogen gas is often used to displace air, the chemist looks up "nitrogen" and "air" and finds the characters for both. These appear in the text in the same phrase as the word *gasu*.

The next ideographs are found after 0°C. The chemist looks up "cooling bath" and finds the ideograph for "bath" is one of those in question. The other turns out to be "ice." The characters found for "slowly" and "introduce" match those in the text. Methyl bromide was used to initiate the reaction. The word "methyl" was identified through the Katakana; words "bromide," "initiate," and "reaction" by looking them up in the dictionary and finding the proper characters.

The following step was complicated by the use of Hirakana characters. To the chemist familiar with Grignard reactions it was apparent that the reaction was continued until the magnesium was practically all dissolved. The supernatent solution was poured off carefully. In a four-necked flask, 430 g of I was placed, the air displaced by nitrogen (same as above). For the synthesis, the ether solution of (II; X = Cl) was added slowly (?) from a separatory funnel with continued stirring (?) keeping the reaction mixture at 20°C. The words "synthesis," "solution," "supernatent," and "separatory funnel" are ideographs recognizable in the glossary. The reaction solution was filtered and the filtrate distilled (characters from glossary). The distillate boiling at $132 \sim 134°C$ is composed of 177 g of III. Yield is 79%, and 165 g of I (bp 160°C) is recovered.

The chemist may also benefit by working out transliterations of Chinese ideographs from words which appear in the glossary. Knowing the transliterations, he can look up words which begin with common ideographs. This can be of considerable help with key words in characters.

The process is tedious and requires patience and some ingenuity, but it can be made to give results. With practice, transliteration of Katakana becomes more rapid and at times the words can be identified by the first one or two characters. Ideographs remembered from one article help with the next. Normally, the chemist would not go into such detail. The article examined turned out to be particularly easy, but it is not likely that there are many from which nothing can be obtained.

A dictionary of some seven thousand technical and semi-technical words published by the Japanese Ministry of Education is particularly helpful. The words are in two lists, one alphabetized according to English and the other according to transliterated Japanese.* The glossary is in two parts, the first for transliterated Katakana, and the second, alphabetized in English, is for ideographs. The words in the glossary which follow are taken from this dictionary by permission of the Japanese Ministry of Education.

* For further information, contact Seifu Kankobutsu Service Center, 2–1 Kasumigaseki, Chiyoda-ku, Tokyo, Japan.

JAPANESE SYLLABARY:

H = Hirakana   K = Katakana

	H	K		H	K		H	K		H	K		H	K
a	あ	ア	i	い	イ	u	う	ウ	e	え	エ	o	お	オ
ka	か	カ	ki	き	キ	ku	く	ク	ke	け	ケ	ko	こ	コ
sa	さ	サ	shi	し	シ	su	す	ス	se	せ	セ	so	そ	ソ
ta	た	タ	chi	ち	チ	tsu	つ	ツ	te	て	テ	to	と	ト
na	な	ナ	ni	に	ニ	nu	ぬ	ヌ	ne	ね	ネ	no	の	ノ
ha	は	ハ	hi	ひ	ヒ	fu	ふ	フ	he	へ	ヘ	ho	ほ	ホ
ma	ま	マ	mi	み	ミ	mu	む	ム	me	め	メ	mo	も	モ
ya	や	ヤ				yu	ゆ	ユ				yo	よ	ヨ
ra	ら	ラ	ri	り	リ	ru	る	ル	re	れ	レ	ro	ろ	ロ
wa	わ	ワ										wo	を	ヲ
ga	が	ガ	gi	ぎ	ギ	gu	ぐ	グ	ge	げ	ゲ	go	ご	ゴ
za	ざ	ザ	ji	じ	ジ	zu	ず	ズ	ze	ぜ	ゼ	zo	ぞ	ゾ
da	だ	ダ	ji	ぢ	ヂ	zu	づ	ヅ	de	で	デ	do	ど	ド
ba	ば	バ	bi	び	ビ	bu	ぶ	ブ	be	べ	ベ	bo	ぼ	ボ
pa	ぱ	パ	pi	ぴ	ピ	pu	ぷ	プ	pe	ぺ	ペ	po	ぽ	ポ
kya	きゃ	キャ				kyu	きゅ	キュ				kyo	きょ	キョ
sha	しゃ	シャ				shu	しゅ	シュ				sho	しょ	ショ
cha	ちゃ	チャ				chu	ちゅ	チュ				cho	ちょ	チョ
nya	にゃ	ニャ				nyu	にゅ	ニュ				nyo	にょ	ニョ
hya	ひゃ	ヒャ				hyu	ひゅ	ヒュ				hyo	ひょ	ヒョ
mya	みゃ	ミャ				myu	みゅ	ミュ				myo	みょ	ミョ
rya	りゃ	リャ				ryu	りゅ	リュ				ryo	りょ	リョ
gya	ぎゃ	ギャ				gyu	ぎゅ	ギュ				gyo	ぎょ	ギョ
ja	じゃ	ジャ				ju	じゅ	ジュ				jo	じょ	ジョ
bya	びゃ	ビャ				byu	びゅ	ビュ				byo	びょ	ビョ
pya	ぴゃ	ピャ				pyu	ぴゅ	ピュ				pyo	ぴょ	ピョ
n	ん	ン												

TRANSLITERATED KATAKANA

Of transliterated Katakana words approximately thirty percent are easily recognizable by their resemblance to English or German. Many more will become so if it is remembered that:

1) The letter *r* may be replaced by *l*.
2) The letter *h* may be replaced by *ph* or *f*.
3) The letter *t* may be replaced by *th*.
4) The letter *j* may be replaced by *d*.
5) The letter *b* may be replaced by *v*.
6) The letter ch may be replaced by *t*.
7) The vowel *u* (and on occasion others) may be omitted.
8) The terminal sound *a* may be understood as *er*.

Examples:

Transliterated Japanese	English or German
adaputâ	adapter
biniru	vinyl
nafutarin	naphthalene
jiazo	diazo

A NOTE ABOUT CHINESE

In Chinese chemical articles, as in those in other languages, chemical formulas and numerals are printed as in English. Just from these it is possible for the chemist familiar with the subject to obtain information, perhaps in considerable amounts. There is nothing corresponding to the Katakana alphabet. However, English words are sometimes inserted to explain certain ideographs and summaries in English are frequently added to Chinese articles.

Appendices

Appendix A — Latin Languages

FRENCH, SPANISH, ITALIAN

LIST OF ELEMENTS

Symbol	French	Spanish	Italian
C	carbone	carbono	carbonio
H	hydrogène	hidrogeno	idrogeno
O	oxygène	oxígeno	ossigeno
N	azote	nitrógeno	azòto
S	soufre	azufre	zolfo (solfo)
P	phosphore	fósforo	fosforo
Cl	chlore	cloro	cloro
Br	brome	bromo	bromo
I	iode	yodo	iodio
F	fluor	flúor	fluoro
Li	lithium	litio	litio
Na	sodium	sodio	sodio
K	potassium	potasio	potassio
Ca	calcium	calcio	calcio
Ba	baryum	bario	bario
Si	silicium	silicio	silicio
Zn	zinc	cinc (zinc)	zinco
Mg	magnésium	magnesio	magnesio
Mn	manganèse	manganeso	manganese
Fe	fer	hierro	ferro
Ni	nickel	níquel	nichel
Cu	cuivre	cobre	rame
Co	cobalt	cobalto	cobalto
Al	aluminium	aluminio	alluminio
Sn	étain	estafío	stagno
Pb	plomb	plomo	piombo

Glossary

French	English	Italian	Spanish
accroître	enlarge	crescere	aumentar
âcre	acrid	acre	acre
agiter	shake	scuotere	sacudir
aigre	sour	agro	agrio
aigu	sharp	affilato	afilado
aiguille	needle	ago	aguja
amende, fin (adj.)	fine	fino	fino
anhydre	anhydrous	anidro	anhidro
apparaître	appear	apparire	aparecer
approximatif	approximate	approssimativo	aproximado
arrêter	stop	arrestare	pararse
augmenter	increase	aumentare	aumentar
bas	low	basso	bajo
bleu	blue	azzurro	azul
bouillir	boil	bollire	hervir
boulette	pellet	pallina	píldora
brûler	burn	bruciare	quemar
brun	brown	marrone	pardo
but	purpose	scopo	intención
cesser	stop	terminare	terminar, cesar
changer	change	cambiare	cambiar
chauffer	warm (v.)	riscaldare	calentar
clair	clear	chiaro	claro
clarifier	clarify	chiarire	clarificar
collectionner (receuillir)	collect	collezionare (raccogliere)	coger
commencer	begin	cominciare	comenzar, empezar
composé	compound	composto	compuesto
congélation	freezing (n.)	congelamento	congelación
convenir	agree	essere d'accordo	estar de acuerdo
corriger	correct	correggere	corregir
couche	layer	strato	estrato
couvrir	cover	coprire	tapar
croûte	crust	crosta	costra

French	*English*	*Italian*	*Spanish*
de	from	da	desde
décrire	describe	descrivere	describir
dégager	give off	liberare	emitir
déplacer	displace	spostare	desplazar
déshydrater	dehydrate	disidratare	deshidratar
détruire	destroy	distruggere	destruir
douteux	doubtful	dubbioso	dudoso
doux	sweet	dolce	dulce
doucement	gently	dolcemente	suavemente
écumer	foam (v.)	schiumare	espumar
épaissir	thicken, cake (v.)	indurire	espesar
épreuve	test	prova	prueba
extraire	draw off	estrarre	extraer
faible	weak	debole	débil
fermer	close (v.)	chiudere	cerrar
feuille	leaf	foglia	hoja
flocon	flake	fiocco	copo
fondre	melt (v.)	fondere	derretirse
fort	strong	forte	fuerte
four	oven	forno	horno
frais	fresh	fresco	fresco
fréquemment	frequently	spesso	frecuentemente
geler	freeze	gelare	congelar
glacial	freezing (a.)	glaciale	glacial
goudron	tar	catrame	brea
goutte à goutte	dropwise	goccia a goccia	gota a gota
goutter	drop (v.)	gocciolare	gotear
gris	grey	grigio	gris
haut	high	alto	alto
heure	hour	ora	hora
idée	idea, concept	idea, concetto	concepto
inchangé	unchanged	invariato	inalterado
jaune	yellow	giallo	amarillo
lent	slow	lento	lento
lessive, soude caustique	lye	liscivia, soda caustica	lejía

French	*English*	*Italian*	*Spanish*
mélanger	mix (v.)	mescolare	mezclar
mésurer	measure (v.)	misurare	medir
mince	thin	sottile	delgado
montrer	show	mostrare	mostrar
moyen	medium	mezzo, modo, mediano	mediano
noir	black	nero	negro
pâte	paste	pasta	pasta
peu	little (adv.)	poco	poco
peu à peu	gradually	gradatamente	gradualmente
plat	dish	piatto	plato
poudre	powder	polvere	polvo
presque	almost	quasi	casi
preuve	proof	prova	comprobación, prueba
prudent	cautious	prudente	precavido, prudente
rapport	report	rapporto	reporte
rassembler	collect	raccogliere	recoger
recherche	research	ricerca	indagación, investigación
récipient	container	recipiente	envase
reflux	reflux	riflusso	reflujo
refroidir	cool	raffreddare	refrescar
remuer	stir	mescolare	menear, remover
rester	remain	rimanere	quedarse
rouge	red	rosso	rojo
sans	without	senza	sin
sécher	dry (v.)	asciugare	secarse
soigneusement	carefully	accuratamente	cuidadosamente
souffler	blow	soffiare	soplar
sujet	subjected	soggetto	súbdito
surface	surface	superficie	superficie
travailler	work	lavorare	trabajar
verre	glass	vetro	vidrio
vers le bas	downward	in giù	hacia abajo
vers le haut	upwards	in sù	arriba
verser	pour	versare	verter
vide	empty	vuoto	vacío

ROMANIAN

LIST OF ELEMENTS

Symbol	Romanian
C	carbon
H	hidrogen
O	oxigen
N	nitrogen
S	sulf
P	fosfor
Cl	clor
Br	brom
I	iod
F	fluor
Li	litiu
Na	sodiu
K	potasiu
Ca	calciu
Ba	bariu
Si	siliciu
Zn	zinc
Mg	magneziu
Mn	mangan
Fe	fier
Ni	nikel
Co	cobalt
Cu	cupru
Al	aluminiu
Sn	staniu
Pb	plumb

Glossary

Romanian	English	Romanian	English
abia	scarcely	bicarbonàt	bicarbonate
absorbi	absorb	brom	bromine
ac	needle	calciu	calcium
accelerà	accelerate	cald	warm (adj.)
acesta	this	càldut	lukewarm
acetic	acetic	cantitate	quantity
acid	acid	carbón	carbon
acidulà	acidify	care	which
acoperì	cover (v.)	caustic	caustic
acru	sour	cercetare	research
acţiune	action	clor	chlorine
activ	active	coace	cake (v.)
adaptà	adapt	colecta	collect (v.)
adăugà	add	colora	dye
adiţional	additional	combinà	combine (v.)
aer	air	complectà	complete (v.)
afinitate	affinity	compus	compound
agità	agitate	comun	common
aînchide	turn off	concentrà	concentrate (v.)
ajuta	help (v.)	condensà	condense
alb	white	conduce	conduct (v.)
albàstru	blue	congela	freeze (v.)
aluminiu	aluminum	considerabil	considerable
amestec	mixture	contamina	contaminate
amesteca	mix (v.)	continuà	continue
apà	water	controlà	control (v.)
apos	aqueous	convenì	agree
aproape	nearly	corecta	correct (v.)
aproximativ	about	creste	increase (v.)
arata	show (v.)	crustă	crust
árde	burn (v.)	culoare	color
ascutit	sharp	cupru	copper
atent	cautious	curs	course
bariu	barium	decànta	decant

Romanian	English	Romanian	English
decolora	decolorize	filtru	filter
dehidrogena	dehydrogenate	flocon	flake (n.)
dela	of, from	folositor	useful
densitate	density	fosfor	phosphorus
descompune	decompose	fracţionà	fractionate (v.)
descreştere	decrease	frecvent	frequently
determinare	determination	galben	yellow
dializă	dialysis	general	general
digerà	digest (v.)	glacial	glacial
diluà	dilute (v.)	gol	empty
diluat	weak	grad	degree
disocia	dissociate	gradat	gradually
disolva	dissolve	granular	granular
dispersà	disperse	greu	difficult
distilà	distil	greutate	weight
distruge	destroy	grijuliu	carefully
dubios	doubtful	gris	grey
dulce	sweet	gros	thick
după	after	hidrogén	hydrogen
eliminà	expel	hidrogenare	hydrogenation
esenţă	essence	hidrolizá	hydrolysis
etuva	oven	homogen	homogeneous
evapora	evaporate	hydrocarburà	hydrocarbon
evità	avoid	identificare	identify
evolùţie	evolution	imediat	immediately
exclude	exclude	împroşca	spatter
éxicator	desiccator	impur	impure
experimènt	experiment	în	in
extern	external	inalt	high
extràge	extract (v.)	începe	begin
face	make (v.)	încetişor	slowly
fără	without	închide	close (v.)
farfurie	dish, plate	închis	dark
fier	iron	incolor	colorless
fierbe	boil (v.)	indeparta	remove
fierbinte	hot	inel	ring (n.)

Romanian	English	Romanian	English
inert	inert	optic	optical
inflamabil	inflammable	ora	hour
influenţà	influence	oxigén	oxygen
insolubil	insoluble	parte	part
intern	internal	pasta	paste (n.)
interval	interval	penetra	permeate
întrebuinţa	use (v.)	periodic	periodic
întrebuinţare	use (n.)	picatura	drop (n.)
introduce	introduce	pierdere	loss
iod	iodine	pîlnie	funnel (n.)
izola	isolate	plasa	place (v.)
jos	low	posibil	possible
larg	large	potasiu	potassium
libera	liberate	precipita	precipitate (v.)
lichefia	deliquesce (v.)	preparare	preparation
limpezi	clarify	presiune	pressure
liquid	liquid (n.)	prezentà	presence
magnéziu	magnesium	prin	through
major	major	proaspăt	fresh
mangán	manganese	proba	test (v.)
maron	brown	procedurà	procedure
màsurà	measurement	proces	process (n.)
mentine	maintain	pudra	powder
mic	small	pulveriza	pulverize
minor	minor	punct de congelare	freezing point
miros	odor	punct de fierbere	boiling point
modificare	modification	punct de topire	melting point
nefolositor	useless	pur	pure
nesaturat	unsaturated	purifica	purify
neutraliza	neutralize	purificare	purification
nichel	nickel	puternic	strong
nitric	nitric	răcì	cool (v.)
nitrogèn	nitrogen	rămîne	remain (v.)
observa	observe	rearanja	rearrange
obtine	obtain	rece	cold
opri	stop (v.)	recipient	container

Romanian	English	Romanian	English
reflux	reflux	strat	layer
regenerare	regeneration	structurà	structure
relata	relate	sub	under
reţine	retain	subiect	subject
rezidiu	residue	sublima	sublimate
rezolva	solve	substanţà	substance
rezultat	result	substitui	substitute (v.)
roşu	red	subtire	thin
rota	rotate	suficient	sufficiently
sarcinà	task	sulf	sulfur
saturat	saturate	suprafaţa	surface
schimbà	change (v.)	suspensie	suspension
scindare	cleave	timp	time
scutura	shake	topi	melt
sediment	sediment	tot	all
sfîrşit	end	transformare	transformation
similar	similar	trata	treat (v.)
sodá	soda	tub	tube
sodiu	sodium	turbiditate	turbidity
solid	solid	turna	pour (v.)
solidifica	solidify	umezi	moisten
solubil	soluble	uree	urea
solubilitate	solubility	uşca	dry (v.)
soluţie	solution	usor	mild
solvent	solvent	uşurel	gently
spàla	wash (v.)	uzual	usually
spontan	spontaneous	vacum	vacuum
spuma	foam (v.)	valvà	valve
sta	stand (v.)	vapori	vapors
stabil	stable	vàs	flask
stabilitate	stability	verde	green
sticlă	glass	vîscous	viscous

Appendix B — Germanic Languages

GERMAN, DUTCH, SWEDISH, DANISH

LIST OF ELEMENTS

Symbol	German	Dutch	Swedish	Danish
C	Kohlenstoff	koolstof	kol	kulstof
H	Wasserstoff	waterstof	väte	brint
O	Sauerstoff	zuurstof	syre	ilt
N	Stickstoff	stikstof	kväve	kvælstof
S	Schwefel	zwavel	svavel	svovl
P	Phosphor	phosphorus	fosfor	fosfor
Cl	Chlor	chloor	klor	klor
Br	Brom	broom	brom	brom
I	Jod	jodium	jod	jod
F	Fluor	fluor	fluor	fluor
Li	Lithium	lithium	litium	litium
Na	Natrium	natrium	natrium	natrium
K	Kalium	kalium	kalium	kalium
Ca	Kalzium	calcium	kalcium	kalcium
Ba	Barium	barium	baryum	barium
Si	Silizium	silicium	kisel	silicium
Zn	Zink	zink	zink	zink
Mg	Magnesium	magnesium	magnesium	magnesium
Mn	Mangan	mangaan	mangan	mangan
Fe	Eisen	ijzer	järn	jern
Ni	Nickel	nikkel	nickel	nickel
Co	Kobalt (Cobaltum)	kobalt	kobolt	kobolt
Cu	Kupfer	koper	kopper	kobber
Al	Aluminium	aluminium	aluminium	aluminium
Sn	Zinn	tin	tenn	tin
Pb	Blei	lood	bly	bly

Glossary

German	English	Dutch	Swedish	Danish
ab	from	van	från, av	fra, af
Abänderung	modification	modificatie	modifikation	modifikation
abblasen	blow off	aflaten	utblåsa	udblæse
abdämpfen, konzentrieren	concentrate	indampen	koncentrera	koncentrere
abgeben	give off	afgeven	avgiva	afgive, udløse
abhelfen	correct	corrigeren	rätta	rette
abkühlen	cool off	afkoelen	svalna, avkyla	afkøle
ablassen	decant	afgieten	avgjuta	afhælde
abnehmen	decrease	verminderen	förminska	aftage, formindske
absatzweise	intermittently	stapsgewijs	stötvist	stødvist
absaugen	draw off (by suction)	afzuigen	avsuga	afsuge
abscheiden	separate	afscheiden	avskilja	adskille
absetzen	deposit	sedimenteren	utskilja	udskille
abstellen	turn off	afsluiten	stänga av	dreje af
abziehen	draw off	afvloeien	avdraka	udtage
Abzug	fume hood	zuurkast	dragskåp	stinkskab
ähnlich	similar	gelijkend	likartet	lignende
allgemein	general	algemeen	allmän	almindelig
allmählich	gradually	langzaam	gradvis	gradvis
ändern	change	veranderen	förandra	forandre
anfangen	begin	beginnen	börja	begynde
anhalten	stop	ophouden	stoppa	stoppe
anhaltend	continuously	voortdurend	kontinuerlig	stadigt
anreichern	enrich, concentrate	concentreren	koncentrera	koncentrere
ansäuren	acidify	aanzuren	syra	gøre sur, forsure
anschliessen	connect	aansluiten	förena	forbinde
Anwesenheit	presence	aanwezigheid	närvaro	tilstedeværelse
arbeiten	work	werken	arbeta	arbejde
auffangen	catch (liquid), collect	opvangen	samla	fange
Aufgabe	problem	probleem	problem	problem
aufhellen	clarify	opheldesen	avklara	afklare
aufhören	discontinue	ophouden	upphöra	holde op

German	English	Dutch	Swedish	Danish
aufrechthalten	maintain	gelijk houden	uppehålla	vedligeholde
aufrühren	stir	roeren	röra	omrøre
Aufschlämmung	suspension	suspensie	uppslämning	opslæmning
auftreten	appear	optreden	visa sig	vise sig
aufwärts	upwards	opwaards	uppåt	opad
Ausbeute	yield	opbrengst	utbytte	udbytte
ausfallen	precipitate	neerslaan	utfälning	bundfald
ausschliessen	exclude	buitensluiten	exkludera	udelukke
aussen	outside of	buiten	yttre	udenfor
austreiben	expel	verdrijven	driva ut	uddrive
austrocknen	dry	uitdrogen	törra	tørre
ausziehen	extract (v.)	extraheren	extrahera	uddrage
Auszug	extract (n.)	uittreksel	extrakt	ekstrakt
Bad	bath	bad	bad	bad
Bau	structure	structuur	struktur	struktur
bedenklich	doubtful	twijfelachtig	tvivelaktig	tvivlsom
bedeutsam	significant	betekenisvol	betydlig	betydelig
befeuchten	moisten	bevochtigen	fukta	fugte
Begriff	idea	begrip	ide	ide
behandeln	treat	behandelen	behandla	behandle
behutsam	cautious	voorzichtig	varsam	forsigtig
beissend	acrid	bijtend	skarp	skarp
benutzen	use (v.), utilize	gebruiken	bruka	bruge
beobachten	observe	waarnemen	observera	iagttage
Bericht	report (n.)	mededeling	rapport	beretning
beschreiben	describe	beschrijven	beskriva	beskrive
besprechen	discuss	bespreken	diskutera	omtale, diskutere
Beständigkeit	stability	bestendigheid	stabilitet	stabilitet
bestimmen	determine	bepalen	bestemma	bestemme
Beweis	proof	bewijs	prov på	bevis
bilden	form	vormen	forma	forme
Blätter	leaves (n.)	blaadjes	blad	blade
blau	blue	blauw	blå	blå
bleiben	remain	blijven	återstå	blive tilbage
brauchbar	useful	bruikbaar	anvendbar	brugbar
braun	brown	bruin	brun	brun

German	English	Dutch	Swedish	Danish
Brei	paste	pasta	pasta	pasta
brennen	burn	branden	brenna	brænde
brüchig	brittle	broos	skör	skør
Dampf	vapor	damp	ånga	damp
darstellen	prepare	prepareren	preparera	preparere
decken	cover	bedekken	täcka	dække
Dichtigkeit	density	dichtheid	densitet	vægtfylde
digerieren	digest	digereren	digestera	digerere
Druck	pressure	druk	tryck	tryk
dunkel	dark	donker	mörk	mørk
dünn	thin	dun	tunn	tynd
durchsetzen	permeate	doordringen	gennomtränga	gennemtrænge
Eigenschaft	property	eigenschap	egenskab	egenskab
eindampfen	boil down, dry by evaporation	verdampen	indunsta	inddampe
einheitlich	homogeneous	homogeen	homogen	homogen
eintragen	introduce	toevoegen	introducera	tilsætte
einwirken	act upon	inwerken	reagere med	reagere med
energisch	energetic	heftig	energisk	energisk
entfärben	decolorize	ontkleuren	avfärga	affarve
entfernen	remove	verwijderen	eliminera	fjerne
entwässern	dehydrate	dehydrateren	dehydrera	afvande
entzündlich	inflammable	brandbaar	lättantändlig	letantændelig
Erfolg, Resultat	result	resultaat	resultat	resultat
erhalten	obtain	verkrijgen	uppnå	opnå
erkalten	cool	afkoelen	svalna	afkøle
erstarren	solidify	stijf worden	göra fast	størkne
extrahieren	extract	extraheren	extrahera	udtrække
Farbe	color	kleur	färg	farve
farblos	colorless	kleurloos	färglös	farveløs
Farbstoff	dye	kleurstof	färgämne	farvestof
fast	almost	bijna	nästan	næsten
fein	fine	fijn	fin	fin
fest	solid	vast	fast	fast
feucht	damp	vochtig	fuktig	fugtig
filtrieren	filter (v.)	filtreren	filtrera	filtrere

German	English	Dutch	Swedish	Danish
Flocke	flake	vlok	flag	fnug
flüssig	liquid	vloeibaar	flytande	flydende
Forschung	research	onderzoek	forsök	forskning
fortsetzen	continue	doorgaan	fortsätta	fortsætte
frieren	freeze	bevriezen	frysa	fryse
frierend	freezing	vriezend	frysande	frysende
frisch	fresh	vers	färsk	frisk
ganz	entirely	geheel	hel	hel
Gefäss	container	vat	behållar	beholder
gelb	yellow	geel	gul	gul
gelinde	gently	zacht	blid	mildt
Gemisch	mixture	mengsel	blanding	blanding
genug	enough	genoeg	nog	nok
gering	small	gering	liten	lille
Geruch	odor	geur	lukt	lugt
geruchlos	odorless	reukloos	luktfri	lugtløs
getrennt	separated	apart	separat	fra hinanden
Gewicht	weight	gewicht	vikt	vægt
gewöhnlich	usually	gewoonlijk	vanligt	almindeligt
giessen	pour	gieten	hälla	hælde
Glas	glass	glas	glass	glas
gleichzeitig	simultaneously	gelijktijdig	samtidigt	samtidigt
Grad	degree	graad	grad	grad
grau	grey	grijs	grå	grå
grün	green	groen	grön	grøn
Harz	resin	hars	harts	harpiks
häufig	frequent	veelvuldig	ofta	ofte
heizen	heat	verhitten	(upp)varma	(op)varme
hell	bright	licht	ljus	lys
hinab	downward	naar beneden	nedåt	nedad
hoch	high	hoog	hög	høj
innerlich	internal	inwendig	inre	indre
klein	small	klein	liten	lille
knapp	close, narrow, tight	nauw	när	nær
Kruste	crust	korst	skorpa	skorpe
kochen	boil	koken	koka	koge

German	English	Dutch	Swedish	Danish
Kochpunkt	boiling point	kookpunt	kokepunkt	kogepunkt
Kohlenwasserstoff	hydrocarbon	koolwaterstof	kolväte	kulbrinte
Kuchen	cake (n.)	koek	kake	kage
kühlen	cool	koelen	svalna	(af)køle
langsam	slow	langzaam	långsam	langsom
Lauge	lye	loog	lut	lud
leer	empty	leeg	tom(t)	tom(t)
löslich	soluble	oplosbaar	upplösbar	opløselig(t)
Löslichkeit	solubility	oplosbaarheid	upplösbarhet	opløslighed
Lösung	solution	oplossing	upplösning	opløsning
Luft	air	lucht	luft	luft
Menge	quantity	hoeveelheid	mängde	mængde
messen	measure	meten	mäta	måle
mischbar	miscible	mengbaar	blandbar	blandelig
mischen	mix	mengen	blanda	blande
Mittel	medium	middel	medel	middel
möglich	possible	mogelijk	möjlig	mulig
nachweisen	show, prove, identify	aantonen	identifiera	identificere
Nädelchen	needles, crystallic	naaldjes	nåle	nåle
Niederschlag	precipitate	neerslag	(ut)fällning	bundfald
niedrig	low	laag	lav	lav
nutzlos	useless	waardeloos	onyttig	ubrugelig
Oberfläche	surface	oppervlakte	yta	overflade
Ofen	oven	oven	ugn	ovn
ohne	without	zonder	utan	uden
Plättchen	platelet, small plate	plaatje	liten plata	lille plade
Probe	test	proef	prov	prøve
Pulver	powder	poeder	pulver	pulver
rasch	quickly	vlug	snappt	hurtigt
rauchend	fuming	rokend	rykande	rygende
regelmässig	regular	regelmatig	regelmässig	regelmæssig
rein	pure	zuiver	ren	ren
reinigen	purify	reinigen	rena	rense
richtig	correct	juist	riktig	rigtig
rot	red	rood	röd	rød

German	English	Dutch	Swedish	Danish
Rückfluss	reflux	terugvloei	återflöde	tilbageløb
Rückstand	residue	overblijfsel	återstod	rest
rühren	stir	roeren	(om)röra	omrøre
sammeln	collect	verzamelen	samla	samle
sanft	soft	zacht	blöd	blød
sättigen	saturate	verzadigen	mätta	mætte
sauer	sour	zuur	sur	sur
Säure	acid	zuur	syra	syre
scharf	sharp	scherp	skarp	skarp
schäumen	(de)foam, skim	schuimen	skumma	skumme
scheiden	separate	afscheiden	åtskilja	adskille
Schicht	layer	laag	lag(et)	lag
Schluss	conclusion	conclusie	slut(en)	slutning
schmelzen	melt	smelten	smälta	smelte
Schmelzpunkt	melting point	smeltpunt	smeltepunkt	smeltepunkt
schnell	rapidly	snel	snappt	hurtig
schütteln	shake	schudden	skaka	ryste
schwach	weak	zwak	svak	svag
schwarz	black	zwart	svart	sort
schwierig	difficult	moeilijk	svår	vanskelig
sofort	immediately	dadelijk	omedelbar	straks
sorgfältig	careful	zorgvuldig	aktsam	omhyggelig
spezifisch	specific	specifiek	specifik	specifisk
spontan	spontaneous	spontaan	omedelbar	spontan, umiddelbar
stark	strong	sterk	stark	stærk
stehen	stand	staan	stå	stå
Stufe	degree	trap	grad	grad
Stunde	hour	uur	timma	time
süss	sweet	zoet	söt	sød
Teer	tar	teer	tjära	tjære
Teil	part	deel	del	del
Teller	dish	schotel	skål	skål
träg(e)	inert	traag	träg	ureagerende
trennen	separate	scheiden	åtskilla	adskille
trocknen	dry	drogen	(ut)torka	tørre
tröpfeln	drop	druppelen	droppa	dryppe

German	English	Dutch	Swedish	Danish
tropfenweise	dropwise	druppelsgewijs	dropvis	dråbevis
Trübung	turbidity	troebelheid	grumlighet	grumsethed
übereinstimmen	agree	overeen komen	överenstemma med	overenstemme med
ungefähr	approximate	ongeveer	cirka	omtrent
unlöslich	insoluble	onoplosbaar	olöslig	uopløselig
unrein	impure	onzuiver	oren	uren
unterwerfen	subject	onderwerpen	underkasta	underkaste
unverändert	unchanged	onveranderd	oändrad	uændret
Ventil	valve	ventiel	ventil	ventil
verändern	change	veranderen	ändra	ændre
Verbindung	compound	verbinding	förening	forbindelse
verdauen	digest	verteren	smälta mat	fordøje
verdünnen	dilute	verdunnen	utspäda	fortynde
Verdünnung	dilution	verdunning	förtunning	fortynding
verdunsten	evaporate	verdampen	evaporera	fordampe
vereinbar	compatible	verenigbaar	jämnförbar	forenelig(t)
vereinigen	combine	verenigen	kombinera	kombinere
Verfahren	procedure	werkwijze	procedur	fremgangsmåde
vermehren	increase	vermeerderen	växa	tiltage
verringern	decrease	verminderen	förminska	formindske
verschliessen	close	afsluiten	stänga	lukke
Versuch	experiment	proef	experiment	experiment
vollständig	complete	geheel	komplet	fuldstændig
Vorgang	process	verloop	process	proces
vorschlagen	propose	voorstellen	föreslå	foreslå
vorsichtig	carefully	voorzichtig	aktsamt	forsigtigt
wärmen	warm	verwarmen	varm	varm
waschen	wash	wassen	tvätta	vaske
wasserfrei	anhydrous	watervrij	ohydrerat	uhydreret, vandfri
wässerig	aqueous	waterig	vattenholdig	vandholdig
weiss	white	wit	vit	hvid
wenig	little	weinig	liten	lidt
wirksam	active	actief	aktiv	aktiv
zäh	viscous	taai	viskos	viskøs
Zeichen	symbol	symbool	symbol	symbol
zeigen	show	tonen	visa	(frem)vise

German	English	Dutch	Swedish	Danish
Zeit	time	tijd	tid	tid
Zerfliessung	deliquescence	vervloeing	delikvescence	delikvescence
zerreiben	pulverize	fijnwrijven	pulverisera	fintmale
zerstören	destroy	vernietigen	förstöra	ødelægge
Zimmer	room	kamer	rum	rum
zunehmen	increase	toenemen	växa	forøge
Zusatz	addition	toevoeging	tillsätning	tilsætning
Zweck	purpose	doel	syfte	formål

Appendix C — Languages Unrelated to English

HUNGARIAN

Symbol	Hungarian
C	szén
H	hidrogén
O	oxigén
N	nitrogén
S	kén
P	foszfor
Cl	klór
Br	bróm
I	jód
F	fluor
Li	litium
Na	nátrium
K	kálium
Ca	kalcium, mész
Ba	bárium
Si	szilicium
Zn	cink
Mg	magnézium
Mn	mangán
Fe	vas
Ni	nikkel
Co	kobalt
Cu	réz
Al	aluminium
Sn	ón
Pb	ólom

Glossary

Hungarian	English	Hungarian	English
ad	give	energikus	energetic
aktív	active	eredmény	result
alacsony	low	erős	strong
alak	form (n.)	értéktelen	useless
alávet	subject (v.)	eszme	idea
áll	stand (v.)	fagyaszt	freeze
anyag	substance	fagyás pont	freezing point
bárium	barium	fanyar	acrid
barna	brown	fehér	white
befed	cover (v.)	fekete	black
belső	internal	felfelé	upwards
besűrít	concentrate (v.)	felület	surface
bevezet	introduce	fenntart	maintain
bizmút	bismuth	fényes	bright
bizonyiték	proof	festék	dye
bróm	bromine	finom	fine
cél	purpose	fok	degree
csapadék	precipitate (n.)	fokozatosan	gradually
csökken	decrease	folyadék	liquid
csöpp	drop	folyamat	process
csöppenként	dropwise	folytat	continue
dekantál	decant	folytonosan	continuously
édes	sweet	forma	form
éget	burn	forral	boil
egyesit	combine	forrás pont	boiling point
egyidejűleg	simultaneously	főz	cook
elég	enough	friss	fresh
elegyíthető	miscible	fürdő	bath
éles	sharp	füstölgő	fuming
eljárás	procedure	gondolat	thought
elkülönit	separate (v.)	gondos	careful
elpárolog	evaporate	gondosan	carefully
eltávolit	remove	gőz	vapor, steam
emészt	digest	gyakori	frequent

Hungarian	English	Hungarian	English
gyanta	resin	kén	sulfur
gyönge	weak	kéreg	crust
gyorsan	quickly	készit	prepare
gyüjt	collect	kétséges	doubtful
gyulékony	inflammable	kever	stir
hab	foam	keverék	mixture
hasonló	similar	kezd	begin
használat	use	kezel	treat (v.)
hasznavehetetlen	useless	kicsiny	small
hasznos	useful	kifujat	blow off
hatástalan	inert	kisérlet	experiment (n.)
héj	crust	kiűz	expel
helyes	correct	kivon	extract (v.)
higany	mercury	kivonat	extract (n.)
higít	dilute (v.)	kizár	exclude
higitás	dilution	koncentrál	concentrate (v.)
hő	heat	következtetés	conclusion
hőmérséklet	temperature	közeg	medium
homogén	homogeneous	közel	close
hozzáadás	addition	külön	separate (adj.)
hűt	cool	különleges	special
idő	time	külső	external
illat	scent	kutatás	research
inditványoz	propose	lapocska	platelet
intézkedik	act upon	lassú	slow
jelenlét	presence	látszik	appear
jelent	report (v.)	lead	give off
jelentékeny	significant	lebocsát	draw off
jelentés	report (n.)	lefelé	downward
jelkép	symbol	lehető	possible
jód	iodine	lehetséges	possible
kap	obtain	lehűt	cool off
kálium	potassium	lehúz	draw off
kályha	oven	leir	describe
kátrány	tar	lerakódás	deposit (n.)
kék	blue	levegő	air

Hungarian	English	Hungarian	English
lezár	turn off	oldhatatlan	insoluble
lúg	lye	oldható	soluble
magas	high	oldhatóság	solubility
magnézium	magnesium	ólom	lead
majdnem	almost	olvad	melt
mangán	manganese	olvadás pont	melting point
marad	remain	önkéntes	spontaneous
maradék	residue	önt	pour
mérték	measure	óra	hour
megáll	stop (v.)	összeáll	coalesce
megfigyel	observe	összeférheto	compatible
meghatároz	determine	összegyüjt	collect
megközelitő	approximate	összeköt	connect
megsemmisít	destroy	összesül	cake (v.)
megszakítva	intermittently	összevet	compare
megszilárdul	solidify	óvatos	cautious
megvitat	discuss	oxigén	oxygen
meleg	warm (adj.)	pép	paste
melegít	heat (v.)	pehely	flake
mennyiség	quantity	por	powder
merev	rigid	porít	pulverize
módositás	modification	próba	test
mos	wash (v.)	probléma	problem
munka	work (n.)	ráz	shake
nátrium	sodium	rendszerint	usually
nedvesít	moisten	rész	part
nehéz	difficult	réteg	layer
nélkül	without	réz	copper
nikkel	nickel	sárga	yellow
nitrogén	nitrogen	sav	acid
növel	increase	savanyít	acidify
nyer	obtain	savanyú	sour
nyeredék	yield	sötét	dark
nyirkos	damp	stabilitás	stability
nyomás	pressure	súly	weight
oldat	solution	suruség	density

Hungarian	English	Hungarian	English
szag	odor	teljesen	entirely
szagtalan	odorless	térfogat	volume
száraz	dry	tevékeny	active
szén	carbon	tiszta	pure
szénhidrogén	hydrocarbon	tisztátlan	impure
szabályos	regular	tisztáz	clarify
szándék	purpose	tisztít	purify
szelep	valve	töményít	concentrate (v.)
szeliden	gently	törékeny	brittle
szellőző fülke	fume hood	tű	needle
szerint	according to	tulajdonság	property
szerkezet	structure	üres	empty
szétfolyó	deliquescent	üveg	glass
szilárd	solid	változás	change
szimbólum	symbol	változatlan	unchanged
szín	color	vas	iron
színtelen	colorless	vegyít	combine
színtelenit	decolorize	vegyület	compound
szoba	room	vékony	thin
szokás	custom	viszkózus	viscous
szokásos	regular	vizes	aqueous
szűr	filter	viztelen	anhydrous
szürke	grey	viztelenít	dehydrate
tál	dish	visszafolyat	reflux
tartály	container	vörös	red
tartósság	stability	wolfram	tungsten
tehetetlen	inert	zavarosság	turbidity
telit	saturate	zöld	green
teljes	complete		

FINNISH

LIST OF ELEMENTS

Symbol	Finnish
C	hiili
H	vety
O	happi
N	typpi
S	rikki
P	fosfori
Cl	kloori
Br	bromi
I	jodi
F	fluori
Na	natriumi
K	kalium
Ca	kalsiumi
Ba	barium
Si	piialkuaine
Zn	sinkki
Mg	magnesiumi
Mn	mangaani
Fe	rauta
Ni	nikkeli
Co	koboltti
Cu	kupari
Al	aluminiumi
Sn	tina
Pb	lyijy

Glossary

Finnish	English	Finnish	English
absorboida	absorb	havaita	observe
aihe	subject (n.)	heikko	weak
aika	time	hidas	slow
aine	substance	hieno	fine
alaspäin	downwards	hienontaa	grind (v.)
alkaa	begin	hiili	carbon
alla	under	hiilivety	hydrocarbon
apu	help (n.)	hitaasti	slowly
asettaa	place (v.)	hiutale	flake (n.)
aste	degree	homogeeninen	homogeneous
asteittain	gradually	huolellinen	careful
astia	container	huolellisesti	carefully
auttaa	help (v.)	hydrata	hydrogenate
bikarbonaatti	bicarbonate	hydrolyysi	hydrolysis
dekantoida	decant	hyödyllinen	useful
dialyysi	dialysis	hyödytön	useless
ehdoton	absolute	hämmentää	stir
eksikaattori	desiccator	hävittää	destroy
emäksinen	alkaline	höyry	vapor
epäpuhdas	impure	ilma	air
eristää	isolate	ilman	without
erotella	fractionate, break down, decompose	jatkaa	continue
		jauhe	powder
erottaa	separate	johtaa	conduct (v.)
etikka-	acetic, (vinegar)	joka	which
haalea	lukewarm	jälkeen	after
haihduttaa	evaporate	jää	ice
hajoittaa	disperse	jäädyttää	freeze
haju	odor	jäädä	remain
hajuton	odorless	jäähdyttää	cool (v.)
halkeaminen	gap	jäännös	residue
happi	oxygen	jäätymispiste	freezing point
happo	acid	kaikki	all
harmaa '	grey	kappale	piece

Finnish	English	Finnish	English
katkera	acrid, bitter	liukoinen	soluble
kehitys	development	liukoisuus	solubility
keltainen	yellow	liuos	solution
kerros	layer	liuotin	solvent
kerrostuma	deposit (n.)	liuottaa	dissolve
kerätä	collect	lopettaa	complete (v.)
kiehua	boil (v.)	loppu	end (n.)
kiehumispiste	boiling point	lyhennelmä	abbreviation (n.)
kiihdyttää	accelerate	lähes	almost
kiinteä	solid (adj.)	lämmin	warm (adj.)
kirkastaa	clarify	lämmittää	warm (v.)
koe	experiment	lämmityslaite	heater
koko	whole	lämpö	heat, temperature
korkea	high	läsnäolo	presence
korvata	substitute (v.)	mahdollinen	possible
korvike	substitute (n.)	makea	sweet
kostea	damp	matala	low
kostuttaa	moisten	melkein	nearly
kuiva	dry (adj.)	melkoinen	considerable
kuivata	dry (v.)	mittaus	measurement
kuori	crust, shell	muunnos	modification
kuuma	hot	muurahaishappo	formic acid
kylliksi	enough	muutos	change (n.)
kylmä	cold	määritys	determination
käsitellä	treat	määrä	quantity
käyttö	use (n.)	neste	liquid (n.)
laimentaa	dilute (v.)	nestemäinen	liquid (adj.)
lasi	glass	neulasia	needles
lievä	mild	neutraloida	neutralize
lievästi	gently	noin	about
liittää	connect	nopeasti	quickly
lipeä	lye	näyte	sample (n.)
lisä-	additional	näyttää	show (v.)
lisäksi	further, in addition to	ohut	thin
lisääntyä	increase (v.)	oikea	correct
liukenematon	insoluble	olennainen	essential, inherent

Finnish	English	Finnish	English
omaisuus	property	rengas	ring
ominaispaino	specific gravity	rikki	sulfur
optillinen	optical	roiskia	spatter (v.)
osa	part (n.)	ruskea	brown
paakkuuntua	cake (v.)	saada	obtain
paine	pressure	saastuttaa	contaminate
paino	weight	sakka	sediment
paksu	thick	samanlainen	similar
palaa	burn (v.)	sameus	turbidity
pallo	pellet	sammuttaa	**turn off, extinguish**
passiivinen	inert	saostaa	precipitate (v.)
peittää	cover (v.)	sauva	rod
pestä	wash (v.)	seisoa	stand (v.)
pidättää	retain	sekoittaa	agitate, stir, blend
pieni	small	sekoitus	mixture
pimeä	dark	sininen	blue
pinta	surface	sisäinen	internal
pisara	drop (n.)	sisältö	content
poistaa	remove	sooda	soda
poistaa väri	discolor	sopia	agree
prosentti	percentage	sovittaa	adapt
prosessi	process (n.)	spontaani	spontaneous
puhdas	pure	sublimoida	sublimate
puhdistaa	purify	sulamispiste	melting point
puhdistus	purification	sulattaa	melt (v.)
pullo	flask	sulkea	close (v.)
punainen	red	sulkea pois	exclude
putki	tube	suodatin	filter (n.)
pystyjäähdytys	reflux	suppilo	funnel
pysyvyys	stability	suspensio	suspension
pysäyttää	stop (v.)	suunta	direction
pyöriä	rotate	suuri	large
rakeinen	granular	syntyvä	nascent
rakenne	structure	säännöllinen	regular
ravistaa	shake	tahmea	viscous
regeneraatio	regeneration	tahna	paste (n.)

Finnish	*English*	*Finnish*	*English*
tarpeeksi	sufficiently	vahva	strong
tavallisesti	usually	vaikea	difficult
tehdä	make	vaikuttaa	influence (v.)
tehdä happameksi	acidify	vakuumi	vacuum
tehtävä	task	valkoinen	white
terävä	sharp	valmistaminen	preparation
tiheys	density	valvoa	control (v.)
tiivistää	concentrate (v.)	valvonta	control (n.)
tislata	distil	vapauttaa	liberate
todistus	proof	varovainen	cautious
toiminta	action	vastaan	against
toimiva	active	venttiili	valve
tuhlaus	waste (n.)	vesi	water
tulenarka	inflammable	vetinen	aqueous
tulos	result (n.)	vetistyä	liquefy, melt
tulppa	stopper	vihreä	green
tunkeutua	penetrate	viileä	cool (adj.)
tunnistaa	identify	virrata	stream (v.)
tunti	hour	virta	stream (n.)
tuoda	bring into	voittaa	gain (v.)
tuore	fresh	voitto	gain (n.)
tuottaa	yield (v.)	vähentää	decrease (v.)
tutkimus	research	väliaika	interval
tyhjä	empty	välittömästi	immediately
typpi-	nitric	välttää	avoid
tyydyttämätön	unsaturated	väri	color (n.)
työ	work (n.)	väritön	colorless
tämä	this	yhdiste	compound
tärkeä	important	yhdistää	combine
täydellinen	complete (adj.)	yhteinen	common
ulkopuolinen	external	yleinen	general
usein	frequently	yli	above, over
uuni	oven	ylijäämä	residue
uuttaa	extract (v.)	ylläpitää	maintain
vaahdota	foam (v.)	ylöspäin	upward
vaahto	foam (n.)		

POLISH, CZECH

LIST OF ELEMENTS

Symbol	Polish	Czech
C	węgiel	uhlík
H	wodór	vodík
O	tlen	kyslík
N	azot	dusík
S	siarka	síra
P	fosfór	fosfor
Cl	chlor	chlor
Br	brom	brom
I	jod	jod
F	fluor	fluor
Li	lit	litík
Na	sód	sodík
K	potas	draslík
Ca	wapień	vápník
Ba	bar	barium
Si	krzem	křemen
Zn	cynk	zinek
Mg	magnez	hořčík
Mn	mangan	mangan
Fe	żelazo	železo
Ni	nikiel	nikl
Co	kobalt	kobalt
Cu	miedź	měd'
Al	glin	hliník
Sn	cyna	cín
Pb	ołów	olovo

Glossary

Polish	Czech	English
abstrahować	abstrahovat	abstract (v.)
aktywny	činný, aktivní	active
alkaliczny	louh	basic
azotowy	dusičný	nitric
badanie	výzkum	research
barwa	barva	color
bez	bez	without
bezbarwny	bez barvy	colorless
bezpośrednio	bezprostředně	immediately
bezwonny	bez vůně	odorless
bialy	bílý	white
blisko	blízko	near, nearly
bronzowy	hnědý	brown
caly	celý	whole
chlodzić	ochladiti	cool (v.)
ciało, substancja	podstata	body, substance
ciemny	temný	dark
cienki	tenký	thin
cieply	teplý	warm
ciężar	váha	weight
ciężar gatunkowy	specifická váha	specific gravity
ciśnienie	tlak	pressure
czas	čas	time
czerwony	červený	red
część	část	part
często	často	frequently
czynić	činiti	do (v.)
czysty	čistý	pure
dalej	další	further
dehydrogenować	dehydrogenovati	dehydrogenate
delikatnie	jemně	gently
dializa	dialysa	dialysis
dodać	přidati	add
dodawanie	přídavek (dodatečně)	addition (in addition)

Polish	Czech	English
dostateczny	dostatečný	sufficient
doswiadczać	vyzkoušeti	experiment (v.)
doświadczenie	pokus	experiment (n.)
dowód	zkouška, důkaz	proof
dwuwęglan	dvojuhličitan	bicarbonate
dystylować	destilovati	distil
działanie	pohyb	action
farbować	barviti	dye (v.)
filtrować	procediti	filter (v.), let through
gęstość	hustota	density
godzina	hodina	hour
gorący	horký	hot
gotować	vařiti	boil
gromadzić	sbírati	collect
gruby	hustý	thick
grzać	hřát	heat (v.)
grzelnik	ohřivač	heater
homogeniczny	stejnorodý	homogeneous
hydrogenizacja	hydrogenace	hydrogenation
hydroliza	hydrolysa	hydrolysis
i, takźe, oraz	a, také, taky	and, also, too
igła	jehla	needle
ilość	množství	quantity
istota	podstata	essence
kaustyczny	žíravý	caustic
klapa	záklopka	valve
kolba	láhev	flask
kontrolować	kontrolovati	control (v.)
kropla	kapka	drop (n.)
który	který	which
kulka	kulička	pellet
kurs	běh	course
kwas	kyselina	acid
kwasić	proměniti v kyselinu	acidify
ledwie	sotva	scarcely
lejek	nálevka	funnel (n.)

Polish	Czech	English
lepki	lepkavý	viscous
lód	led	ice
lodowaty	ledový	glacial
łączyć	spojovati	combine
mały	malý	small
marznąć	zmraziti	freeze
mętność	zkalenost	turbidity
mierzenie	měření	measurement
mieszać	míchati	stir
mieszanina	směs	mixture
mniejszy	menší	minor
mocny	silný	strong
mocznik	močovina	urea
modyfikacja	pozměnění	modification
możliwy	možný	possible
mrówczany	mravenčí	formic
myć	mýti	wash
na	na	on
naczynie	nádoba	vessel
nad	nahoře	above
następnie	přívrženci	following
nasycać	nasytiti	saturate
niebieski	modrý	blue
nieczysty	nečistý	impure
nienasycony	nenasycený	unsaturated
nierozpuszczalny	nerozpustný	insoluble
niski	nízký	low
objętość	obsah	contents
obracać	otáčeti	rotate
obszerny	velký	large
octowy	octový	acetic
oczyszczác	očistiti	purify
oczyszczenie	očištění	purification
odbarwić	odbarviti	decolorize
oddzielic	odloučiti	isolate
odrodzenie	obroditi	regeneration

Polish	Czech	English
odstep	mezera	interval
odwoderować	zbaviti vodíku	dehydrogenate
ogrzać	rozpáliti	warm (v.)
okazać	objasniti	show (v.)
około	okolo	about
okréslenie	určení	determination
opowiadać	líčiti, sděliti	relate
optyczny	optický	optical
osad	usazenina	sediment
ostrożny	opatrný	cautious
ostry	ostrý	sharp
osusznik	vysušovač	desiccator
palić	spáliti	burn (v.)
palny	hořlavý	flammable
para	pára	vapor
pasta	pasta	paste (n.)
patka	prut	rod
perjodyczny	periodický	periodic
piec	pec	oven
płatek	vločka	flake (n.)
płyn	tekutina	liquid (n.)
płynny	tekutý	liquid (adj.)
po	po, za	after
pod	pod, v	under, in
podobny	podobný	similar
pokrewienstwo	příbuzenství	affinity
pokryć	pokrývati	cover (v.)
ponad	do, nahoru	up, upwards
poprawiać	opraviti	correct (v.)
potok	proud	stream
powierzchnia	plocha	surface
powietrze	vzduch	air
powolny	zdlouhavý	slow
powszechny	obecný	general
pozostałość	usazenina	residue
pozostawać	zůstati	remain

Polish	*Czech*	*English*
pożyteczny	užitečný	useful
prawie	skoro	almost
próbka	vzorek	sample
procedura	postup	procedure
procent	procento	percent
proszek	prach	powder
prowadzić	vésti	conduct (v.)
próżnia	vzduchoprázdný, vakuum	vacuum
próżny	prázdný	empty
przeciw	proti	against
przegrupować	znova uspořádati	rearrange
przeksztalcenie	přeměna	transformation
przenikać	proniknouti	permeate
przez	přes, skrze	through
przygotowanie	příprava	preparation
przyspieszyć	urychliti	accelerate
przystosować	přizpůsobiti	adapt
punkt topienia	bod tání	melting point
punkt wrzenia	bod varu	boiling point
punkt zamarzania	bod mrazu	freezing point
regularny	pravidelný	regular
rodzący się	vznikající	nascent
rozcieńczyć	rozřediti	dilute (v.)
rozdrabniać	rozlomiti	fractionate
rozgrzać	rozehřáti	warm (v.)
rozkładać	rozložiti	decompose
rozłączyć	rozděliti	dissociate
rozproszyć	rozptýliti	disperse
rozpuszczalność	rozpustnost	solubility
rozpuszczalny	rozpustný	soluble
rura	trubička (zkumavka)	tube (test tube)
siwy	šedý	grey
skalać	poskvrniti	contaminate
skorupa	kůra	crust
słaby	slabý	weak
słodki	sladký	sweet

Polish	Czech	English
soda	soda	soda
spontanicrny	samovolný	spontaneous
stałosć	stálost	stability
stanąć	státi	stand (v.)
starannie	opatrně, pečlivě	carefully
stopniowo	postupně	gradually
strata	ztráta	loss
struktura	struktura	structure
suszyć	sušiti	dry (v.)
świeżo	svěží	freshly
szkło	sklo	glass
szybko	rychle	quickly
ten	tento	this
topnieć	taviti	melt
trudny	těžký, nesnadný	difficult
trwać	zůstati, vytrvati	persist
trwaly	neměnný	stable
twardy	pevný	solid
ubywać	ubývati	decrease
umieścić	postaviti	place (v.)
unikać	vyhýbati se	avoid
usunąć	odkliditi	remove
utożsamić	určovati	identify
utrzymywać	udržovati	maintain
uzyskać	dosáhnouti	obtain
używać	užívati	use (v.)
warstwa	vrstva	layer
wątpliwy	pochybný	doubtful
wchłaniać	vsáti	absorb
węglowodór	uhlovodík	hydrocarbon
wilgotny	vlhký	damp
własność	vlastnost	property
woda	voda	water
wodny	vodnatý	aqueous
wolno	pomalu	slowly
wspólny	společný	common
wstrząsać	třásti	shake

Polish	*Czech*	*English*
wszystek	celý, všechen	all
wyłączyć	vyloučiti	exclude
wyparować	vypařiti	evaporate
wypędzić	vyhnati	drive out
wysoki	vysoký	high
wystarczająco	dostatečně	sufficiently
wysuszony	suchý	dry (adj.)
wyzwalać	osvoboditi, uvolniti	liberate
wzrastać	růsti, vzrůstati	increase (v.), grow (v.)
z	od, z	of, from, with
zaczynać	začíti	begin
zadanie	úloha	task
zakończenie	konec	end
zakończyć	ukončiti	close (v.)
zamykać	vypnouti	turn off
zapach	vůně, **zápach**	odor
zasadowy	zásaditý	basic
zatkać	zastaviti	stop (v.)
zatrzymywać	zadržeti	retain
zawieszenie	závěs, suspense	suspension
zazwyczaj	obyčejně	usually
zbiornik	nádoba	container
zgadzać się	shodnouti se	agree
zgęszczać	zhustiti	condense
ziarnisty	zrnitý	granular
zielony	zelený	green
zimny	studený	cold
zmieniać	měniti	change (v.)
znaczny	nemalý	considerable
znak przeczący	záporný znak	negative sign
zniszczyć	**zničiti**	destroy
zobojętniać	neutralisovati	neutralize
zupełny	úplný	complete
związek	sloučenina	compound (n.)
zwilżyć	navlhčiti	moisten

Appendix D — Russian

LIST OF ELEMENTS

Symbol	Russian	Transliterated Russian
C	углерод	uglerod
H	водород	vodorod
O	кислород	kislorod
N	азот	azot
S	сера	sera
P	фосфор	fosfor
Cl	хлор	khlor
Br	бром	brom
I	иод	iod
Li	литий	litii
Na	натрий	natrii
K	калий	kalii
Ca	кальций	kal'tsii
Ba	барий	barii
Si	силиций	silitsii
Zn	цинк	tsink
Mg	магний	magnii
Mn	марганец	marganets
Fe	железо	zhelezo
Ni	никель	nikel
Cu	медь	med
Al	алюминий	alyuminii
Sn	олово	olovo
Pb	свинец	svinets

Russian Chemical Words

Russian	Transliterated	English
метил	metil	methyl
этил	etil	ethyl
пропил	propil	propyl
изопропил	izopropil	isopropyl
бутил	butil	butyl
изобутил	izobutil	isobutyl
амил	amil	amyl
гексил	geksil	hexyl
гептил	geptil	heptyl
октил	oktil	octyl
нонил	nonil	nonyl
децил	detsil	decyl
цетил	tsetil	cetyl
метанол	metanol	methanol
этанол	etanol	ethanol
метил меркаптан	metil merkaptan	methyl mercaptan
метиламин	metilamin	methylamine
диметиламин	dimetilamin	dimethylamine
триметиламин	trimetilamin	trimethylamine
муравьиная кислота	murav'inaya kislota	formic acid
уксусная кислота	uksusnaya kislota	acetic acid
фенил	fenil	phenyl
бензол	benzol	benzene
бензил	benzil	benzyl
хлорбензол	khlorbenzol	chlorobenzene
дихлорбензол	dikhlorbenzol	dichlorobenzene
бромбензол	brombenzol	bromobenzene
нитробензол	nitrobenzol	nitrobenzene
фенол	fenol	phenol
тиофенол	tiofenol	thiophenol
пирокатехин	pirokatekhin	pyrocatechol
резорцин	rezortsin	resorcinol
гидрохинон	gidrokhinon	hydroquinone
нафтил	naftil	naphthyl

Russian	Transliterated	English
нафтол	naftol	naphthol
нафтиламин	naftilamin	naphthylamine
бензойная кислота	benzoinaya kislota	benzoic acid
бензонитрил	benzonitril	benzonitrile
дифенил метил-	difenil metil-	diphenyl methyl
трифенил метил-	trifenil metil-	triphenyl methyl
фталевая кислота	ftalevaya kislota	phthalic acid
изофталевая кислота	izoftalevaya kislota	isophthalic acid
терефталевая кислота	tereftalevaya kislota	terephthalic acid
хлористоводородная кислота	khloristovodorodnaya kislota	hydrochloric acid
азотная кислота	azotnaya kislota	nitric acid
серная кислота	sernaya kislota	sulfuric acid
вода	voda	water
сернистый водород	sernistyi vodorod	hydrogen sulfide
аммиак	ammiak	ammonia
хлористый аммоний	khloristyi ammonii	ammonium chloride
гидроокись натрия	gidrookis natriya	sodium hydroxide
гидроокись калия	gidrookis kaliya	potassium hydroxide
карбонат натрия	karbonat natriya	sodium carbonate
карбонат калия	karbonat kaliya	potassium carbonate
нитрат натрия	nitrat natriya	sodium nitrate
нитрат калия	nitrat kaliya	potassium nitrate
хлористое железо	khloristoe zhelezo	ferric chloride
хлористый алюминий	khloristyi alyuminii	aluminum chloride

RUSSIAN INTERNATIONAL WORD LIST

Russian	Transliterated	English
аберрация	aberratsiya	aberration
абсолютный	absolyutnyi	absolute
абсорбировать	absorbirovat	absorb
абсорбция	absorbtsiya	absorption
автоклав	avtoklav	autoclave
автоматически	avtomaticheski	automatically
агитатор	agitator	agitator

Russian	*Transliterated*	*English*
аккумулятор	akkumulyator	accumulator, storage battery
акселератор	akselerator	accelerator
активный	aktivnyi	active
алембик	alembik	alembic
алифатический	alifaticheskii	aliphatic
алкалоид	alkaloid	alkaloid
алкил	alkil	alkyl
алкилизовать	alkilizovat	alkylate
алкоголь	alkogol	alcohol
алюминий	alyuminii	aluminum
амальгама	amal'gama	amalgam
амбра	ambra	amber, resin
аметист	ametist	amethyst
амид	amid	amide
аморфный	amorfnyi	amorphous
амплитуда	amplituda	amplitude
амфотерность	amfoternost	amphoteric character
анализ	analiz	analysis
анализировать	analizirovat	analyze
аналогия	analogiya	analogy
анаэробный	anaerobnyi	anaerobic
ангидрид	angidrid	anhydride
анод	anod	anode
аномальный	anomal'nyi	anomalous
антидот	antidot	antidote
антипод	antipod	antipode
антифриз	antifriz	antifreeze
апертура	apertura	aperture
аппарат	apparat	apparatus
ароматический	aromaticheskii	aromatic
асбест	asbest	asbestos
атмосфера	atmosfera	atmosphere
атом	atom	atom
ацетилирование	atsetilirovanie	acetylation
ациклический	atsiklicheskii	acyclic
аэратор	aerator	aerator

Russian	Transliterated	English
аэробный	aerobnyi	aerobic
баласт	balast	ballast
барка	barka	bark, barge
бассейн	bassein	basin, tank
бомба	bomba	bomb
бром	brom	bromine
бромирование	bromirovanie	bromification
бронза	bronza	bronze
бунзен	bunzen	bunsen
буфер	bufer	buffer
бюретка	byuretka	buret
вагон	vagon	wagon, railroad car
вакансия	vakansiya	vacancy
вакуум	vakuum	vacuum
вакцина	vaktsina	vaccine
валентность	valentnost	valence
вариация	variatsiya	variation
ватерпруф	vaterpruf	waterproof
вафля	vaflya	waffle, wafer
вентилятор	ventilyator	ventilator, fan
вибратор	vibrator	vibrator
витамин	vitamin	vitamin
вольтаж	vol'tazh	voltage
вязкий	vyazkii	viscous
газ	gaz	gas
газгольдер	gazgol'der	gas holder
галид	galid	halide
галлон	gallon	gallon
галоген	galogen	halogen
галоидирование	galoidirovanie	halogenation
гель	gel	gel, jelly
гибрид	gibrid	hybrid
гигроскопический	gigroskopicheskii	hygroscopic
гидрат	gidrat	hydrate
гидролиз	gidroliz	hydrolysis
гипотеза	gipoteza	hypothesis

Russian	*Transliterated*	*English*
гипс	gips	gypsum
гланда	glanda	gland
глицерин	glitserin	glycerol
гомогенизатор	gomogenizator	homogenizer
гомолог	gomolog	homolog
гормон	gormon	hormone
гравиметрический	gravimetricheskii	gravimetric
грамм	gramm	gram
графит	grafit	graphite
группа	gruppa	group
деалкилизовать	dealkilizovat	dealkylate
дегенерат	degenerat	degenerate
дегидратовать	degidratovat	dehydrate
дезинфектант	dezinfektant	disinfectant
декантация	dekantatsiya	decantation
декарбонизация	dekarbonizatsiya	decarbonization
декстро	dekstro	dextro
денатурализация	denaturalizatsiya	denaturalization
денсиметр	densimetr	densimeter, hydrometer
деполяризовать	depolyarizovat	depolarize
депрессия	depressiya	depression
дестиллировать	destillirovat	distil
дефект	defekt	defect
дефлагратор	deflagrator	deflagrator
дефростер	defroster	defroster
диализ	dializ	dialysis
диафрагма	diafragma	diaphragm, membrane
диета	dieta	diet
дилатометр	dilatometr	dilatometer
димер	dimer	dimer
дипольный	dipol'nyi	dipole
диск	disk	disc, plate
диссоциировать	dissotsiirovat	dissociate
доза	doza	dose, portion
драйер	draier	dryer
дренаж	drenazh	drainage

Russian	Transliterated	English
жакет	zhaket	jacket
желатин	zhelatin	gelatin
журнал	zhurnal	journal
зона	zona	zone, area
зонд	zond	sound, probe
идеально	ideal'no	ideally
идентифицировать	identifitsirovat	identify
изолировать	izolirovat	isolate
изомеризовать	izomerizovat	isomerize
изотоп	izotop	isotope
имид	imid	imide
ингибитор	ingibitor	inhibitor
ингредиент	ingredient	ingredient
индекс	indeks	index
индикатор	indikator	indicator
инертный	inertnyi	inert
инженер	inzhener	engineer
инкремент	inkrement	increment
инсектисид	insektisid	insecticide
институт	institut	institute
инструмент	instrument	instrument
интегратор	integrator	integrator
интервал	interval	interval
интерес	interes	interest
иод	iod	iodine
ион	ion	ion
исключение	isklyuchenie	exclusion
йодометрический	iodometricheskii	iodometric
кабель	kabel	cable
кавитация	kavitatsiya	cavitation
калий	kalii	potassium
калория	kaloriya	calorie
кальций	kal'tsii	calcium
капилляр	kapillyar	capillary
капсюль	kapsyul	capsule
карта	karta	card, chart

Russian	Transliterated	English
катализатор	katalizator	catalyst
катион	kation	cation
каустический	kausticheskii	caustic
керосин	kerosin	kerosene
кетон	keton	ketone
класс	klass	class
кливаж	klivazh	cleavage
кокс	koks	coke
коллоид	kolloid	colloid
колориметр	kolorimetr	colorimeter
комплекс	kompleks	complex
конденсация	kondensatsiya	condensation
конденсировать	kondensirovat	condense
контакт	kontakt	contact
контекст	kontekst	context
контроль	kontrol	control
конфигурация	konfiguratsiya	configuration
концентрировать	kontsentrirovat	concentrate
концепция	kontseptsiya	conception
координата	koordinata	coordinate
коррозия	korroziya	corrosion
коэфициент	koefitsient	coefficient
крекинг	kreking	cracking
крип	krip	creep
кристаллизовать	kristallizovat	crystallize
кэк	kek	cake (v.)
лаборатория	laboratoriya	laboratory
лак	lak	lacquer
лакриматор	lakrimator	lacrimator
лактаза	laktaza	lactose
латекс	lateks	latex
левый	levyi	levo-, left
лимит	limit	limit
литература	literatura	literature
логарифм	logarifm	logarithm
магний	magnii	magnesium

Russian	Transliterated	English
максимум	maksimum	maximum
манипуляция	manipulyatsiya	manipulation
манускрипт	manuskript	manuscript
масса	massa	mass
материал	material	material
металл	metall	metal
метеор	meteor	meteor
метод	metod	method
метр	metr	meter
механизм	mekhanizm	mechanism
механик	mekhanik	mechanic
мигрировать	migrirovat	migrate
микроскоп	mikroskop	microscope
минерал	mineral	mineral
минорный	minornyi	minor
минута	minuta	minute
модель	model	model
молекула	molekula	molecule
моль	mol	mole
момент	moment	moment
мономер	monomer	monomer
натрий	natrii	sodium
негатив	negativ	negative
нейтрон	neitron	neutron
непропорциональность	neproportsional'nost	disproportion
никель	nikel	nickel
нипель	nipel	nipple
нитрировать	nitrirovat	nitrate
нормальный	normal'nyi	normal
нос	nos	nose, forepart
овал	oval	oval
озонолиз	ozonoliz	ozonolysis
окисление	okislenie	oxidation
окуляр	okulyar	ocular
олеум	oleum	oleum
оператор	operator	operator

Russian	*Transliterated*	*English*
органический	organicheskii	organic
оригинал	original	original
ориентация	orientatsiya	orientation
осмоз	osmoz	osmosis
памфлет	pamflet	pamphlet
паста	pasta	paste
период	period	period
перпендикуляр	perpendikulyar	perpendicular
пестик	pestik	pestle
пигмент	pigment	pigment
пик	pik	peak
пипетка	pipetka	pipette
пирометр	pirometr	pyrometer
плакат	plakat	placard, poster
план	plan	plan
пластмасса	plastmassa	plastic
плунжер	plunzher	plunger, piston
плюс	plyus	plus
позиция	pozitsiya	position
полимер	polimer	polymer
полимеризация	polimerizatsiya	polymerization
полиморф	polimorf	polymorph
полисульфид	polisul'fid	polysulfide
поляроид	polyaroid	polaroid
пора	pora	pore
порция	portsiya	portion
поташ	potash	potash
потенциометр	potentsiometr	potentiometer
практический	prakticheskii	practical
препарат	preparat	preparation
пресс	press	press
проблема	problema	problem
прогрессивный	progressivnyi	progressive
продукт	produkt	product
пропорциональный	proportsional'nyi	proportional
процент	protsent	percent

Russian	Transliterated	English
процесс	protsess	process
псевдо	psevdo	pseudo
пульс	pul's	pulse
радиация	radiatsiya	radiation
радикал	radikal	radical
ракета	raketa	rocket
реагент	reagent	reagent
реакция	reaktsiya	reaction
реализовать	realizovat	realize
регулятор	regulyator	regulator
редуцировать	redutsirovat	reduce
резюме	rezyume	summary
рельс	rel's	rail, track
рессурс	ressurs	resource
реторта	retorta	retort
рефрактометр	refraktometr	refractometer
рецепт	retsept	recipe, prescription
риск	risk	risk
ритм	ritm	rhythm
риформинг	riforming	reforming
робот	robot	robot, automatic device
роль	rol	role
ромбический	rombicheskii	rhombic
сахар	sakhar	sugar
сегмент	segment	segment
седло	sedlo	saddle, seat
секрет	sekret	secret
секунда	sekunda	second
сентиметр	sentimetr	centimeter
сигнал	signal	signal, sign
символ	simbol	symbol
симметрия	simmetriya	symmetry
симптом	simptom	symptom
синтез	sintez	synthesis
система	sistema	system
сифон	sifon	syphon

Russian	*Transliterated*	*English*
скетч	sketch	sketch
скелет	skelet	skeleton, frame
скорчинг	skorching	(rubber) scorching
скруббер	skrubber	scrubber
сода	soda	soda
сорт	sort	sort, kind
спектр	spektr	spectrum
спираль	spiral	spiral
стабилизатор	stabilizator	stabilizer
стандарт	standart	standard
старт	start	start
стиль	stil	style
структура	struktura	structure
сублимат	sublimat	sublimate
сульфированный	sul'firovannyi	sulfonated
сульфон	sul'fon	sulfone
сумма	summa	sum
сфера	sfera	sphere
схема	skhema	scheme, plan
сцена	stsena	stage, scene
табель	tabel	table, schedule
такса	taksa	tax, fixed price
тальк	tal'k	talc
тара	tara	tare
текст	tekst	text
температура	temperatura	temperature
тент	tent	tent
теория	teoriya	theory
терминология	terminologiya	terminology
термометр	termometr	thermometer
термос	termos	thermos
тест	test	test
технический	tekhnicheskii	technical
тинктура	tinktura	tincture
тип	tip	type
титр	titr	titer

Russian	Transliterated	English
титровать	titrovat	titrate
титул	titul	title
траверс	travers	traverse
траектория	traektoriya	trajectory
транспорт	transport	transport
трансформатор	transformator	transformer
турбидиметрия	turbidimetriya	turbidimetry
турбулентность	turbulentnost	turbulence
унификация	unifikatsiya	unification
уошер	uosher	washer
урна	urna	urn
фаза	faza	phase
факт	fakt	fact
фактис	faktis	factice
фактор	faktor	factor
ферментация	fermentatsiya	fermentation
фибра	fibra	fiber
фибрилла	fibrilla	fibril
фигура	figura	figure
физический	fizicheskii	physical
филлер	filler	filler
фильм	fil'm	film
фильтровать	fil'trovat	filter
финиш	finish	finish
флотация	flotatsiya	flotation
фокус	fokus	focus
фонтан	fontan	fountain
форма	forma	form, shape
формация	formatsiya	formation
формула	formula	formula
фрагмент	fragment	fragment
фракционирование	fraktsionirovanie	fractionation
фронт	front	front
функция	funktsiya	function
хаки	khaki	khaki
характеризовать	kharakterizovat	characterize

Russian	*Transliterated*	*English*
химический	khimicheskii	chemical
хлор	khlor	chlorine
хлорировать	khlorirovat	chlorinate
целит	tselit	celite
целлюлоза	tsellyuloza	cellulose
целлюлоид	tsellyuloid	celluloid
цемент	tsement	cement
центр	tsentr	center
центрифуга	tsentrifuga	centrifuge
цеолит	tseolit	zeolite
цикл	tsikl	cycle
цилидр	tsilidr	cylinder
цинк	tsink	zinc
циркуляр	tsirkulyar	circular
цистерна	tsisterna	cistern, reservoir
цитата	tsitata	citation
шанс	shans	chance
шкала	shkala	scale
шок	shok	shock
штат	shtat	state
эквивалентный	ekvivalentnyi	equivalent
эквилибр	ekvilibr	equilibrium
экзаменовать	ekzamenovat	examine
экономный	ekonomnyi	economical
экспериментальный	eksperimental'nyi	experimental
экстракт	ekstrakt	extract
эксцесс	ekstsess	excess
электролит	elektrolit	electrolyte
электрон	elektron	electron
элемент	element	element
эмблема	emblema	emblem
эмульсия	emul'siya	emulsion
энергичный	energichnyi	energetic
энергия	energiya	energy
энзим	enzim	enzyme
эфир	efir	ether
эффект	effekt	effect

Russian Technical Words from French

Russian	Transliterated	French	English
азот	azot	azote	nitrogen
аллонж	allonzh	allonge	adapter
ампула	ampula	ampoule	ampoule
арматура	armatura	armature	mounting
аффинаж	affinazh	affinage	refining
баллон	ballon	ballon	(gas) cylinder
бетон	beton	béton	concrete
бидон	bidon	bidon	container
блиндаж	blindazh	blindage	screen, shield
брикет	briket	briquette	briquet
брошюра	broshyura	brochure	brochure
вентиль	ventil	ventillon	valve
жавелева вода	zhaveleva voda	eau de Javel	Javel water, bleach
иприт	iprit	ypérite	mustard gas
капот	kapot	capot	hood, housing
каучук	kauchuk	caoutchouc	rubber
кювет	kyuvet	cuvette	cell, vessel
монтаж	montazh	montage	installation
ореол	oreol	auréole	aureole, corona
пудра	pudra	poudre	powder
реактив	reaktiv	réactif	reagent, agent
регламент	reglament	règlement	regulation, rule
резюме	rezyume	résumé	resume, summary
рельеф	rel'ef	relief	relief, contour
рулон	rulon	rouleau	roller, roll
спектр	spektr	spectre	spectrum
тур	tur	tour	turn
фасон	fason	façon	style, fashion
флакон	flakon	flacon	flask
шанс	shans	chance	chance
шифр	shifr	chiffre	cipher, code
эпюр	epyur	épure	diagram, line
этаж	etazh	étage	story, level
эталон	etalon	étalon	standard (of weights)
этап	etap	étape	stage, station

Russian Technical Words from German

Russian	Transliterated	German	English
абзац	abzats	Absatz	paragraph, item
абрис	abris	Abriss	outline, sketch
абцуг	abtsug	Abzug	fume hood
абшайдер	abshaider	Abscheider	separator, trap
аншлаг	anshlag	Anschlag	estimate, attempt
аптека	apteka	Apotheke	drugstore
вагон	vagon	Waggon	railroad car
ванна	vanna	Wanne	tub, vat, tank
вассерглас	vasserglas	Wasserglas	water glass
вата	vata	Watte	wadding
вахта	vakhta	Wacht	watch, duty
вахтер	vakhter	Waechter	watchman
велер	veler	Waehler	selector
вентиль	ventil	Ventil	valve
винт	vint	winden	wind, coil
вольфрам	vol'fram	Wolfram	tungsten
кали	kali	Kali	potash
калий	kalii	Kalium	potassium
керн	kern	Kern	nucleus, core
кизельгур	kizel'gur	Kieselguhr	diatomaceous earth
кнопка	knopka	Knopf	button, knob
кокс	koks	Koks	coke
колба	kolba	Kolben	flask, retort
линза	linza	Linse	lens
линия	liniya	Linie	line, mark
лупа	lupa	Lupe	magnifying glass
люфт	lyuft	Luft	gap, clearance
масштаб	masshtab	Mass-stab	scale, rule
мундштук	mundshtuk	Mundstück	mouthpiece, nozzle
натрий	natrii	Natrium	sodium
нуль	nul	Null	zero
обер-	ober-	Ober-	over-, superior
обертон	oberton	Oberton	overtone
пломба	plomba	Plombe	stamp, seal
пункт	punkt	Punkt	point, spot

Russian	Transliterated	German	English
рама	rama	Rahmen	frame, casing
рейсфедер	reisfeder	Reiss-feder	drawing pen
рейсшина	reisshina	Reiss-schiene	T-square
рейтер	reiter	Reiter	rider (analytical balance)
ремень	remen	Riemen	belt, strap
реферат	referat	Referat	abstract, review
референт	referent	Referent	abstractor
ригель	rigel	Riegel	rail, cross bar
сталь	stal	Stahl	steel
стул	stul	Stuhl	chair, seat
тигель	tigel	Tiegel	crucible
торф	torf	Torf	peat
трегер	treger	Traeger	carrier
тушь	tush	Tusch	India ink
фабрика	fabrika	Fabrik	factory, mill
факел	fakel	Fackel	torch, flare
фирма	firma	Firma	firm, company
флюс	flyus	Fluss	flux
форштосс	forshtoss	Vorstoss	adapter
фрахт	frakht	Fracht	freight
хлор	khlor	Chlor	chlorine
цанги	tsangi	Zange	tongs, forceps
цель	tsel	Ziel	goal, aim
цитата	tsitata	Zitat	quotation
цифра	tsifra	Ziffer	figure, digit
шайба	shaiba	Scheibe	washer, disk
шахта	shakhta	Schacht	mine, pit
шина	shina	Schiene	tire
ширма	shirma	Schirm	screen, shield
шлак	shlak	Schlacke	slag, cinder
шлам	shlam	Schlamm	slime, sludge
шланг	shlang	Schlange	hose
шлиры	shliry	Schlieren	schlieren
шлиф	shlif	Schliff	cut section
шлиц	shlits	Schlitz	slit, slot
шнур	shnur	Schnur	cord, string

Russian	Transliterated	German	English
шпиц	shpits	Spitze	spire, steeple
шприц	shprits	Spritze	injector, sprayer
шрифт	shrift	Schrift	type, character
штамп	shtamp	Stampfe	stamp, punch
штанга	shtanga	Stange	pole, pillar
штраф	shtraf	Strafe	fine, penalty
штрих	shtrikh	Strich	stroke, touch

Glossary

Russian	Transliterated	English
абстракция	abstraktsiya	abstract
азотный	azotnyi	nitric
активный	aktivnyi	active
без	bez	without
без запаха	bez zapakha	odorless
белый	belyi	white
бережно	berezhno	carefully
бесполезный	bespoleznyi	useless
бесцветный	bestsvetnyi	colorless
больший	bol'shii	major
большой	bol'shoi	large
брызги	bryzgi	spatter
было	bylo	nearly
быстро	bystro	quickly
вакуум	vakuum	vacuum
вверх	vverkh	up, upwards
вес	ves	weight
вести	vesti	conduct, lead (v.)
весь	ves	all, whole, entire
вешание	veshanie	suspension
вещество	veshchestvo	substance
взбалтывать	vzbaltyvat	shake
видоизменение	vidoizmenenie	modification
влиять	vliyat	influence

Russian	Transliterated	English
вместилище	vmestilishche	container
внести	vnesti	introduce
внешний	vneshnii	external
внутренний	vnutrennii	internal
вода	voda	water
водяной	vodyanoi	aqueous
возгонять	vozgonyat	sublimate
воздух	vozdukh	air
возможно	vozmozhno	possible
возрастать	vozrastat	increase
возрождение	vozrozhdenie	regeneration
волновать	volnovat	agitate
вращать	vrashchat	rotate
время	vremya	time
всасывать	vsasyvat	absorb
выгонять	vygonyat	expel
выделять	vydelyat	liberate, free, isolate, extract
вынимать	vynimat	extract (v.)
высокий	vysokii	high
вязкий	vyazkii	viscous
гидрирование	gidrirovanie	hydrogenation
гидролиз	gidroliz	hydrolysis
голубой	goluboi	blue, light blue
гореть	goret	burn (v.)
горючий	goryuchii	inflammable
горячий	goryachii	hot
градус	gradus	degree
грелка	grelka	heater
греть	gret	heat (v.)
густота	gustota	density
давать	davat	yield
давление	davlenie	pressure
дальше	dal'she	further
двууглекислый	dvuuglekislyi	bicarbonate
дегидрировать	degidrirovat	dehydrogenate
действие	deistvie	action

Russian	Transliterated	English
делать	delat	make (v.), do
делать прояснение	delat proyasnenie	clarify
диализ	dializ	dialysis
добавлено	dobavleno	additional
добиться	dobit'sya	obtain
довольно	dovol'no	enough, rather
доказательство	dokazatel'stvo	proof
достаточно	dostatochno	sufficiently
духовка	dukhovka	oven
дымоход	dymokhod	funnel (n.)
едва	edva	scarcely
едкий	edkii	caustic
жёлтый	zheltyi	yellow
жидкий	zhidkii	liquid (adj.)
жидкость	zhidkost	liquid (n.), thin
заботливый	zabotlivyi	careful
заведывать	zavedyvat	control
загрязнять	zagryaznyat	contaminate
задание	zadanie	task
закипеть	zakipet	boil, bring to boil
закрывать	zakryvat	close (v.), cover
закрыть	zakryt	turn off
замерзать	zamerzat	freeze
замещение	zameshchenie	substitution
запах	zapakh	odor
запомнить	zapomnit	retain in memory
заслужить	zasluzhit	gain (v.)
затвердевать	zatverdevat	cake (v.), harden
зелёный	zelenyi	green
зернистый	zernistyi	granular
знак минус	znak minus	negative sign
зрительный	zritel'nyi	optical
и, также, тоже	i, takzhe, tozhe	and, also, too
игла	igla	needle
из, от	iz, ot	of, from
избегать	izbegat	avoid
изменение	izmenenie	rearrangement

Russian	*Transliterated*	*English*
изолировать	izolirovat	isolate
инертный	inertnyi	inert
исключать	isklyuchat	exclude
испарение	isparenie	vapor
испарять	isparyat	evaporate
исследование	issledovanie	research
иодная	iodnaya	iodine
капля	kaplya	drop (n.)
кислород	kislorod	oxygen
кислота	kislota	acid
клапан	klapan	valve
клейстер	kleister	paste (n.)
колба	kolba	flask
количество	kolichestvo	quantity
кольцо	kol'tso	ring
конец	konets	end
коричневый	korichnevyi	brown
корка	korka	cork, stopper; crust
краска	kraska	dye, color
красний	krasnii	red
крепкий	krepkii	stable, firm, strong
который	kotoryi	which
ледяной	ledyanoi	glacial
лечить	lechit	treat, cure (medic.)
лёд	led	ice
маленький	malen'kii	small
медленно	medlenno	slowly
медленный	medlennyi	slow
медь	med	copper
мелкий	melkii	fine, small, shallow
меньший	men'shii	minor
мерка	merka	measurement, sign
мочевина	mochevina	urea
муравьиный	murav'inyi	formic
мутность	mutnost	turbidity
мыть	myt	wash
мягкий	myagkii	mild, soft

Russian	*Transliterated*	*English*
мягко	myagko	gently
на	na	on, in, to
наблюдать	nablyudat	observe
нагреть	nagret	heat (v.), warm up
над	nad	above, over
налить	nalit	pour
насыщать	nasyshchat	saturate
находить связь	nakhodit svyaz	relate
начать	nachat	begin
немало	nemalo	considerably
ненасыщенный	nenasyshchennyi	unsaturated
нерастворимый	nerastvorimyi	insoluble
нечистый	nechistyi	impure
низкий	nizkii	low
обесцвечивать	obestsvechivat	decolorize
образ деиствия	obraz deistviya	procedure, method
образчик	obrazchik	sample
общий	obshchii	general, common
обычно	obychno	usually
обычный	obychnyi	regular
объединить	ob'edinit	unite
однородный	odnorodnyi	homogeneous
около	okolo	about, near
опознавать	opoznavat	identify
определение	opredelenie	definition
опыт	opyt	experiment, experience
осадок	osadok	sediment
осаждать	osazhdat	precipitate (v.)
основной	osnovnoi	basic
оставаться	ostavat'sya	remain
остановить	ostanovit	stop (v.)
остаток	ostatok	residue
осторожный	ostorozhnyi	cautious
острый	ostryi	sharp
отдельный	otdel'nyi	separate (adj.)
отлив	otliv	reflux
отнимать	otnimat	subtract (v.)

Russian	*Transliterated*	*English*
очищать	ochishchat	purify
пар	par	steam
пениться	penit'sya	foam (v.)
переварить	perevarit	digest
перегонять	peregonyat	distil
переменить	peremenit	change (v.)
поверхность	poverkhnost	surface
под	pod	under
поддерживать	podderzhivat	maintain, support
подкислять	podkislyat	acidify
подобный	podobnyi	similar
показывать	pokazyvat	show (v.)
покрывать	pokryvat	cover (v.)
полезный	poleznyi	useful
полный	polnyi	complete, full
полный сомнений	polnyi somnenii	doubtful, very dubious
пользоваться	pol'zovat'sya	use (v.), make use
помешать	pomeshat	stir, disturb
помогать	pomogat	help (v.)
поправлять	popravlyat	correct (v.)
порошок	poroshok	powder
после	posle	after
поставить	postavit	place (v.)
постепенно	postepenno	gradually
(по)стоять	(po)stoyat	stand up
почти	pochti	almost
превращение	prevrashchenie	transformation
прибавить	pribavit	add
приготовление	prigotovlenie	preparation
приспособлять	prisposoblyat	adapt
присутствие	prisutstvie	presence
пробка	probka	stopper
проверить	proverit	test (v.), verify
продолжать	prodolzhat	continue (v.)
промежуток	promezhutok	interval
проникать	pronikat	permeate
пропажа	propazha	loss

Russian	*Transliterated*	*English*
против	protiv	against, opposite
процеживать	protsezhivat	filter
процент	protsent	percent
пустой	pustoi	empty
пустыня	pustynya	waste
работа	rabota	work (n.)
развёртывание	razvertyvanie	development
разгонять	razgonyat	disperse
разжижать	razzhizhat	dilute (v.)
разлагать	razlagat	dissociate
разлагаться	razlagat'sya	decompose
разрушить	razrushit	destroy
распространённый	rasprostranennyi	common
раствор	rastvor	solution
растворимость	rastvorimost	solubility
растворимый	rastvorimyi	soluble
растворитель	rastvoritel	solvent
растворять	rastvoryat	dissolve
растворяться	rastvoryat'sya	deliquesce
растирать	rastirat	pulverize
расщепление	rasshcheplenie	splitting, lamination
результат	rezul'tat	result
ректификация	rektifikatsiya	purification, rectification
решать	reshat	resolve (v.)
рождающийся	rozhdayushchiisya	nascent
самопроизвольность	samoproizvol'nost	spontaneity
свежий	svezhii	fresh
свойство	svoistvo	property
сгущать	sgushchat	condense
сера	sera	sulfur
серый	seryi	grey
сильный	sil'nyi	strong
скиснуть	skisnut	sour (v.)
слабый	slabyi	weak
сладкий	sladkii	sweet
следующее	sleduyushchee	following
слой	sloi	layer

Russian	Transliterated	English
смесь	smes	compound
смешать	smeshat	mix (v.)
смешивание	smeshivanie	mixing
собрать	sobrat	collect (v.)
согласиться	soglasit'sya	agree
сода	soda	soda
содержание	soderzhanie	contents
соединить	soedinit	connect
соединять	soedinyat	join
соль	sol	salt
сосредоточить	sosredotochit	concentrate (v.)
сосуд	sosud	vessel
сочетать	sochetat	combine (v.)
способ	sposob	process (n.)
спускающийся	spuskayushchiisya	descending
сразу	srazu	immediately
сродство	srodstvo	affinity
стекло	steklo	glass
стержень	sterzhen	rod
струя	struya	stream
стыть	styt	cool (v.)
сушёный	sushenyi	dry (adj.)
сушилка	sushilka	desiccator
сушить	sushit	dry (v.)
сущность	sushchnost	essence
спеживать	stsezhivat	decant
сырость	syrost	dampness
тарелка	tarelka	dish
таять	tayat	melt (v.)
твердеть	tverdet	solidify
твёрдыи	tverdyi	solid
тело	telo	body
тема	tema	subject
тепло	teplo	warm
тепловатый	teplovatyi	lukewarm
течение	techenie	course
тёмный	temnyi	dark

Russian	*Transliterated*	*English*
толщиной	tolshchinoi	thick
тонкий	tonkii	thin
точка замерзания	tochka zamerzaniya	freezing point
точка кипения	tochka kipeniya	boiling point
точка плавления	tochka plavleniya	melting point
трубочка	trubochka	tube
трудный	trudnyi	difficult
убрать	ubrat	remove (v.)
увлажнять	uvlazhnyat	moisten
углеводород	uglevodorod	hydrocarbon
углерод	uglerod	carbon
удельный вес	udel'nyi ves	density
уксусный	uksusnyi	acetic
уменьшение	umen'shenie	decrease
употребление	upotreblenie	use (n.)
ускорять	uskoryat	accelerate
усреднять	usrednyat	neutralize
устойчивость	ustoichivost	stability
устройство	ustroistvo	structure
фракционировать	fraktsionirovat	fractionate
хлопья	khlop'ya	flakes (n.)
холодно	kholodno	cold
цвет	tsvet	color
целый	tselyi	whole
час	chas	hour
часто	chasto	frequently
часть	chast	part
через	cherez	through, in
чистый	chistyi	pure
шарик	sharik	pellet
щёлок	shchelok	lye
это	eto	this (adj.), this is

Appendix E — Japanese

LIST OF ELEMENTS

Symbol	Transliterated	Japanese
C	tanso	炭素
H	suiso	水素
O	sanso	酸素
N	chisso	窒素
S	iô	硫黄
P	rin	リン
Cl	enso	塩素
Br	shuso	臭素
I	yôso	ヨウ素
F	fusso	フッ素
Li	richûmu*	リチウム
Na	natoryûmu*	ナトリウム
K	karyûmu*	カリウム
Ca	karushûmu*	カルシウム
Ba	baryûmu*	バリウム
Si	keiso	ケイ素
Zn	aen	亜鉛
Mg	maguneshûmu*	マグネシウム
Mn	mangan	マンガン
Fe	tetsu	鉄
Ni	nikkeru	ニッケル
Co	kobaruto	コバルト
Cu	dô	銅
Al	aruminyûmu*	アルミニウム
Sn	suzu	スズ
Pb	namari	鉛

* Words marked with an asterisk differ slightly from the Japanese equivalent but are the accepted standard form in Japanese-English dictionaries.

Japanese Glossary

Japanese	English	Japanese	English
adaputâ	adapter	aseton	acetone
aguromereishon	agglomeration	ashiru	acyl
aji (kabutsu)	azi(de)	asukorubin	ascorbic
ajipin	adipic	asuparagin	asparagine
akurijin	acridine	asupirêtâ	aspirator
akuriru	acrylic	asutachin	astatine
akurorein	acrolein	awa (dachi)	bubble, foam(ing)
amarugamu	amalgam	azo	azo
amido	amide	baioretto	violet
amin	amine	bânâ	burner
amino	amino	banajûmu	vanadium
amiru	amyl	barero-	valero-
ammonia	ammonia	baryûmu*	barium
anirin	aniline	benjijin	benzidine
aisôru	anisole	benjiru	benzyl
anpuru	ampule	benzaru	benzal
antimon	antimony	benzen	benzene
antorakinon	anthraquinone	benzoiru	benzoyl
antoraniru	anthranilic	beriryûmu*	beryllium
antorasen	anthracene	bi-	bi-
aranin	alanine	bîkâ	beaker
arîru	aryl	bin	bottle
ariru	allyl	biniru	vinyl
arudehido	aldehyde	bisu-	bis-
aruginin	arginine	bisukôsu	viscose
arukari	alkali	bisumasu	bismuth
arukaroido	alkaloid	bonbe	bomb
arukiru	alkyl	buchin	butyne
arukôru	alcohol	buchiru	butyl
aruminyûmu*	aluminum	Buhhunâ	Büchner
asechiren	acetylene	burakku	black
asechiru	acetyl	burû	blue
asetâru	acetal	butan	butane
aseto-	aceto-	buten	butene

Japanese	English	Japanese	English
byuretto	burette	gomu	gum (rubber)
chiazôru	thiazole	Gûchi-rutsubo	Gooch crucible
chio-	thio-	gurafaito	graphite
chioniru	thionyl	gurikôru	glycol
chitan	titanium	gurîn	green
chûbu	tube	Guriniyâru	Grignard
deka-	deca-	guriseriru	glyceryl
dekantêshon	decantation	haipo-	hypo-
deshikêtâ	desiccator	hakari	balance, scale
dodeshiru	dodecyl	harogen (kabutu)	halogen (halide)
dorai	dry	hekisa-	hexa-
echiren	ethylene	hekisan	hexane
echiru	ethyl	heputan	heptane
ekisutorakuto	extract	heryûmu*	helium
eoshin	eosine	hidorajin	hydrazine
epikuroru-hidorin	epichlorohydrin	hidorokinon	hydroquinone
epokisido	epoxide	hidorokishiru	hydroxyl
esteru	ester	horumu-	form-
etan	ethane	hosugen	phosgene
etanôru	ethanol	ichi-	mono-
êteru	ether	imino-	imino-
feniru	phenyl	indôru	indole
ferishianido	ferricyanide	iô	sulfur
feroshianido	ferrocyanide	ion	ion
firutâ	filter	iso-	iso-
fukka(butsu)	fluor(ide)	ji-	di-
fumaru	fumaric	jiazo	diazo
furakkusu	flux	jisian	cyanogen
furan	furan	jûteriumu	deuterium
furasuko	flask	juwâ-bin	Dewar vessel
furufurâru	furfural	kâbaido	carbide
furuoren	fluorine	kadomyûmu*	cadmium
futaru	phthalic	kakimaze	stirring, agitation
garasu	glass	kapuro-	capro-
gasu	gas	karubamin	carbamic
geru	gel	karuben	carbene

Japanese	English	Japanese	English
karubinôru	carbinol	maron	malonic
karubokishiru	carboxyl	mechionin	methionine
karuboniru	carbonyl	mechiru	methyl
karushûmu*	calcium	mejichiru	mesityl
karyûmu*	potassium	Mekeru	Meker (burner)
kasei	caustic	menisukasu	meniscus
keihi	cinnamic	merisiru	melissyl
keisôdo	diatomaceous earth (kieselguhr)	merukaputan	mercaptan
kenka	saponify	mesu (furasuko)	measuring (volumetric flask)
keten	ketene	meta	meta
keto-	keto-	metan	methane
keton	ketone	metaroido	metalloid
kiiro	yellow	metokishiru	methoxyl
kinon	quinone	mirisutin	myristic
kinorin	quinoline	moribuden	molybdenum
kisantogen	xanthic	moru	mole
kishiren	xylene	morutaru	mortar
kobaruto	cobalt	myôban	alum
kohaku	succinic	nafuchiru	naphthyl
kondensâ	condenser	nafutarin	naphthalene
konpaundo	compound	natoryûmu*	sodium
koroido	colloid	nega	negative
kôrudo	cold	neo-	neo-
koruku	cork	nikkeru	nickel
kuen	citric	nitoriru	nitrile
kumarin	coumarin	nitoro-	nitro-
kurezôru	cresol	nitoroso-	nitroso-
kuromatogurafî	chromatography	noniru	nonyl
kuromiru	chromyl	oiru	oil
kurômu	chromium, chromic	okishi	hydroxy
kurorohorumu	chloroform	okishimu	oxime
kuroru-	chloro-	okuchiru	octyl
maguneshûmu*	magnesium	okutan	octane
mangan	manganese	orein	olefin, olefinic
marein	maleic	orenji	orange

Japanese	English	Japanese	English
-ôru	-ol	sakkarin	saccharin
oruto	ortho	sarichiru	salicylic
osumyûmu*	osmium	sebashin	sebacic
ôtokurêbu	autoclave	sechiru	cetyl
ozon	ozone	sen	stopper, plug
para	para	seren	selenium
parafin	paraffin	serurôzu	cellulose
parumichin	palmitic	setan	cetane
pentan	pentane	shian	cyanic
pepuchido	peptide	shikuro-	cyclo-
pikurin	picric	shimen	cymene
pimerin	pimelic	shinnamiru	cinnamyl
pinakôru	pinacol	shiran	silane
pinchi-kokku	pinch cock	shirika	silica
pinku	pink	shirikôn	silicon
pipetto	pipette	shirokisan	siloxane
pirijin	pyridine	shisu-	cis-
piro-	pyro-	shisuchin	cystine
pirorijin	pyrrolidine	sô	bath
pirôru	pyrrole	sôda	soda
pirubin	pyruvic	suberin	suberic
pori-	poly-	suchiren	styrene
porufirin	porphyrin	suchiruben	stilbene
puropan	propyl	supâteru	spatula
rakutamu	lactam	supekutoru	spectrum
rakuton	lactone	surari	slurry
raurin	lauric	surufin	sulfinic
reddo	red	suruhon	sulfonic
redokkusu	redox	sutearin	stearic
richûmu*	lithium	sutorippingu	stripping
rin	phosphorus	sutoronchûmu*	strontium
ringo	malic	suzu	tin, stannic
ritomasu	litmus	syû	oxalic
rubijûmu*	rubidium	tangusuten	tungsten
rutsubo	crucible, pot	tanpakushitsu	protein

Japanese	English	Japanese	English
tenbin	balance, scale	toruijin	toluidine
tetora-	tetra-	uran	uranium
toransu-	trans-	yôdohorumu	iodoform
tori-	tri-	yôka	iodide
toruen	toluene	zoru	sol

NUMERALS

one	一	ichi
two	二	ni
three	三	san
four	四	shi (yon)
five	五	go
six	六	roku
seven	七	shichi
eight	八	hachi
nine	九	ku (kyu)
ten	十	jū

Common Ideographs (Chinese)

acetic	酢	saku
acid	酸	san
addition	付　加	huka
alicyclic	脂 環 式	sikansiki
analysis	分　析	bunseki
anhydride	無 水 物	musuibutsu
anion	陰 イ オ ン	in-ion
aqua regia	王　水	ôsui
aromatic	芳 香 族	hôkôzoku
arsenic	ヒ　素	hiso
atom	原　子	genshi
azeotrope	共沸混合物	kyôfutsu-kongôbutsu
base	塩　基	enki
benzoic acid	安 息 香 酸	ansokukôsan
bicarbonate	重 炭 酸 塩	jûtansan'en
bisulfate	重 硫 酸 塩	jûyrûsan'en
boiling	沸　騰	futtô
boron	木 ウ 素	hôso
bromine	臭　素	shûso
carbon	炭　素	tanso
cation	陽 イ オ ン	yô-ion
chemistry	化　学	kagaku
chlorine	塩　素	enso
compound	化 合 物	kagôbutsu
concentration	濃　度	nôdo
condenser	冷 却 器	reikyakuki

English	Japanese	Romaji
copper	銅	dô
crystal	結晶	kesshô
crystallization	晶出	shôshutsu
decomposition	分解	bunkai
density	密度	mitsudo
dilution	希釈度	kishakudo
distillate	留出物	ryûshutsubutsu
distillation	蒸留	zyôryû
dropping	滴	teki
electric	電氣	denki
emulsification	乳化	nyûka
equilibrium	平衡	heikô
extraction	抽出	chûshutsu
filter	沪過器	rokaki
fluorine	フッ素	fusso
fractionating column	精留塔	seiryûtô
fractionation	分別	bunbetsu
freezing point	凝固点	gyôkoten
fulminate	雷酸塩	raisan'en
funnel	漏斗	rôto
gold	金	kin
hydrate	水和物	suiwabutsu
hydro-acid	水素酸	suisosan
hydrogen	水素	suiso
hydrolysis	加水分解	kasui-bunkai
hydroxide	水酸化物	suisankabutsu
iodine	ヨウ素	yôso

English	Japanese	Romaji
iron	鉄	tetsu
isomerism	異性	isei
lactic	乳	nyû
lead	鉛	namari
lipid	脂質	shishitsu
liquid	液体	ekitai
mercury	水銀	suigin
metal	金属	kinzoku
mixture	混合物	kongôbutsu
mother liquor	母液	boeki
neutral	中性	chûsei
nitric	硝	shô
nitrogen	窒素	chisso
normality	規定度	kiteido
oil	油	yu, abura
optical activity	旋光性	senkôsei
organic	有機	yûki
oxygen	酸素	sanso
platinum	白金	hakkin
polarity	極性	kyokusei
precipitation	沈澱	chinden
primary	第一	dai'ichi
protein	蛋白質	tanpakushitsu
radiation	放射	hôsha
reaction	反応	hannô
reagent	試薬	shiyaku
recrystallization	再結晶	saikesshô

reduced pressure	減	圧	gen'atsu
reduction	還	元	kangen
reflux	還	流	kanryû
residue	残	分	zanbun
salt	塩		en
saturation	飽	和	hôwa
secondary	第	二	daini
separation	分	離	bunri
silicon	ケ イ	素	keiso
silver	銀		gin
solution	溶	液	yôeki
steam	(水) 蒸	氣	(sui)jôki
stereo	立	体	rittai
straight chain	直	鎖	chokusa
substitution	置	換	chikan
sulfuric	硫		ryû
tertiary	第	三	daisan
thermometer	温	計	ondokei
thionation	加	硫	karyû
urea	尿	素	nyôso
vacuum	眞	空	shinkû
valeric	吉	草	kissô
vapor phase	氣	相	kisô
viscosity	粘	度	nendo
water	水		sui, mizu
weight	重	量	jûryô
zinc	亜	鉛	aen